SPACE FLIGHT

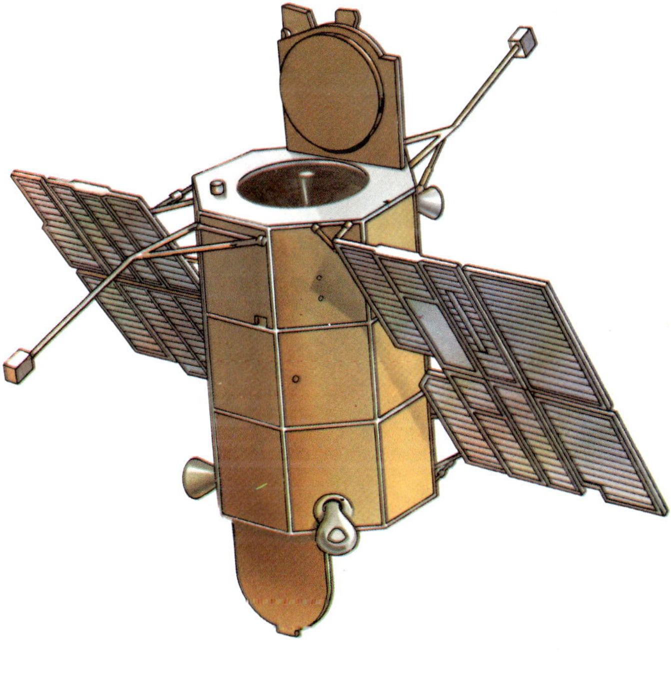

Contents

First Steps into Space	6
Rockets	8
Humans in Space	12
The First Space Station	18
Exploring the Moon	20
The Apollo Program	28
Salyut	44
Skylab	46
Looking Down on Earth	54
Probing the Planets	58
The Space Shuttle	70
More Facts	76
Index	92

Published in 1982 by Rand McNally & Company

First published in 1981 by Pan Books Ltd., London
Designed and produced by
Grisewood & Dempsey Ltd., London

Copyright © by Piper Books Ltd. 1981
U.S. edition © Piper Books Ltd. 1982

Library of Congress Catalog
Card No. 82-60156

Printed by Graficas Reunidas S.A.,
Madrid, Spain
All rights reserved
First printing 1982

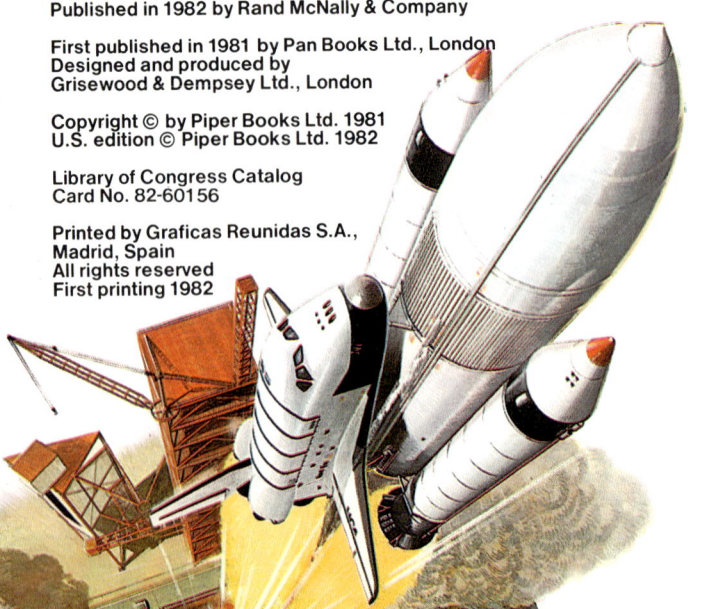

SPACE FLIGHT

Stewart Cowley

Series Design: David Jefferis

RAND McNALLY & COMPANY
Chicago • New York • San Francisco

First Steps into Space

The Space Age began on October 4, 1957. On that day a huge rocket blasted off from the Soviet Union (Russia). *Sputnik 1* was carried in the last section of the rocket. *Sputnik 1* was a small globe of shiny aluminum. It was only 2 feet (60 cm) in diameter. *Sputnik 1* was the first **satellite** made by humans. A satellite is an object that orbits (travels around) a larger object.

Soon *Sputnik 1* was orbiting the earth. It went more than 16,940 miles (27,000 km) per hour. It sent radio signals back to stations far below.

About a month later, the Soviet Union launched (sent out) another satellite. This was *Sputnik 2*. A passenger was on board—the first traveler in space.

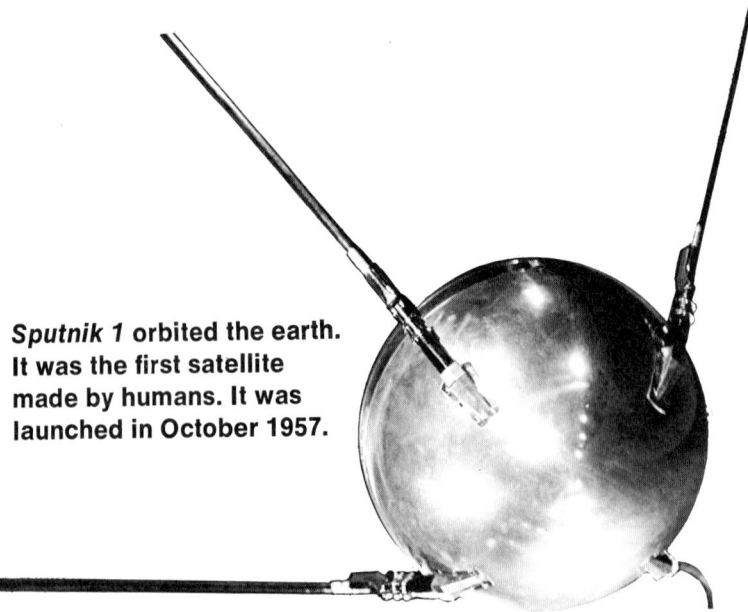

Sputnik 1 orbited the earth. It was the first satellite made by humans. It was launched in October 1957.

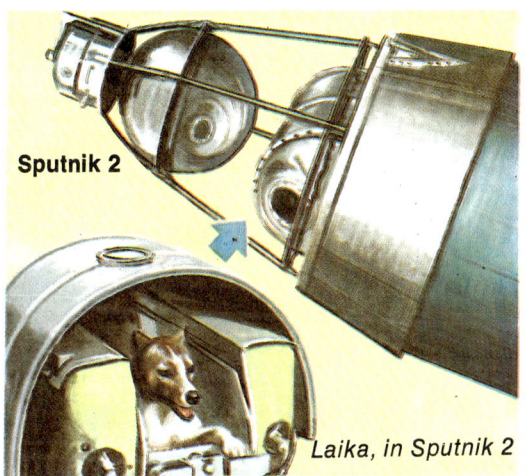
Sputnik 2

Laika, in Sputnik 2

A Dog in Orbit

The first space traveler was Laika, a female dog. She orbited the earth in *Sputnik 2*. She made her trip inside a tiny cabin. The cabin was attached to the last stage of the rocket.

Laika could help herself to food and water from special containers. Instruments gathered information about how she was getting along. This helped scientists learn how people would get along in space. Unfortunately, scientists didn't know how to get Laika back to earth. So she died in space.

Laika was not the first animal to travel in a rocket. Beginning in 1949 many dogs were launched as high as 62 miles (100 km). Scientists wanted to know if animals could travel in rockets. These dogs were brought safely back to earth.

Rockets

Before we can reach outer space, we must escape the pull of **gravity.**

Gravity is the force that pulls objects toward the center of the earth. It makes objects fall to the ground instead of shoot up in the air.

Escaping from gravity takes a lot of speed and power. Scientists began developing rockets powerful enough to reach outer space. The Chinese invented the rocket more than 1,000 years ago. At first, rockets were used in wars. In the 1900s scientists began to find other uses for rockets.

A Law of Nature

A rocket works like a balloon. When the end of a balloon is opened, air rushes out. This makes the balloon

(Above) When a balloon is opened, air rushes out. This makes the balloon fly just as fast as the air. But the balloon goes in the other direction. That's how rockets work, too.

fly just as fast as the air. The balloon is pushed in the opposite direction.

Inside of a rocket gases are heated by fuel. These hot gases rush out. The rocket is pushed in the opposite direction.

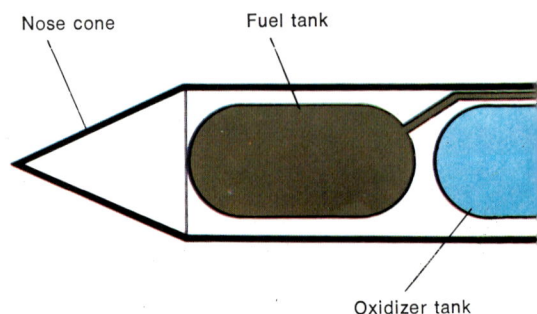

Nose cone Fuel tank

Oxidizer tank

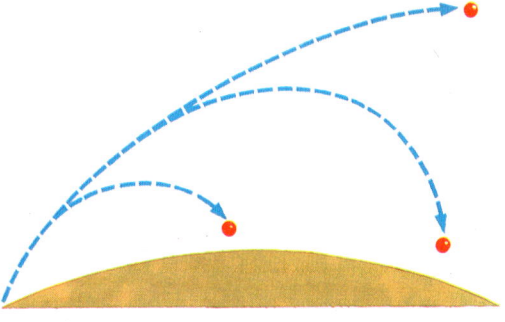

Rockets and balloons work like this because it's a law of nature. This law is called Newton's third law of motion. It says that when energy pushes one way, just as much force pushes in the opposite direction.

To reach outer space a rocket must escape the force of gravity. It must go faster than 18,125 miles (29,000 km) per hour. If it goes too slowly, it will fall back to earth. Huge engines and a lot of fuel are needed.

(Above) An object that leaves the earth at less than 18,125 miles (29,000 km) per hour will fall back to the earth. If it goes 18,125 miles (29,000 km) per hour, it will go into orbit. To leave the earth entirely, it must go 25,000 miles (40,000 km) per hour or more.

(Below) To burn, fuel needs oxygen. Fuel gets it from an *oxidizer*. A rocket pumps fuel and oxidizer into a combustion chamber. There, they mix and burn.

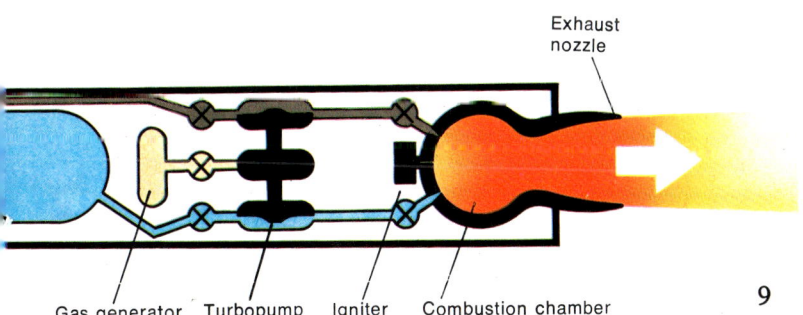

Multistage Rockets

A **stage** is a part of a rocket. Each stage has its own engine and fuel. A rocket that has two or more stages is called a **multistage rocket.** Multistage rockets can go faster than rockets with just one stage.

The United States took some *V-2* rockets from Germany in World War II. A smaller rocket was placed on top of a *V-2*. This smaller rocket was a WAC *Corporal*. When the *V-2* was going as fast as it could, the *Corporal* was fired. The force of the firing increased the speed of the *Corporal*. And the *Corporal* traveled faster without the weight of the *V-2*. The

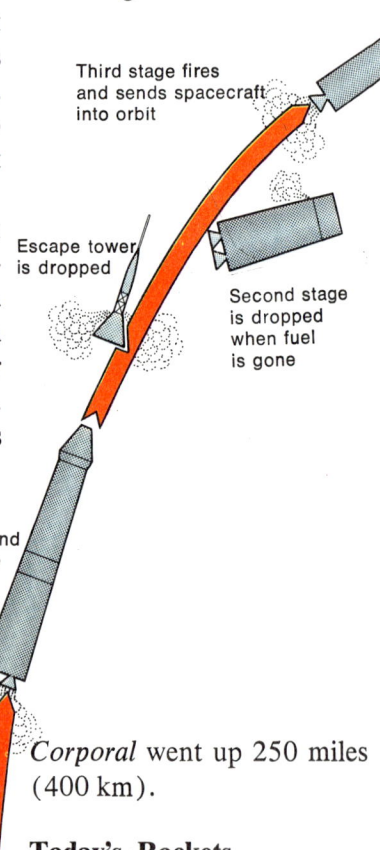

Corporal went up 250 miles (400 km).

Today's Rockets
The **launch vehicle** sends the rocket up. The lighter this vehicle is, the easier it is for the engines to push it up.

Third stage separates

Spacecraft continues in orbit

(Left) To escape gravity, a spacecraft must go very fast. That's why scientists invented multistage rockets. When each stage has used its fuel, it is dropped. This makes the rest of the rocket lighter. Then the next stage fires. The rocket goes faster. (Right) The American *Saturn 5* launch vehicle did a good job. It was used for the *Apollo* moon flights.

That's why most of today's rockets have three stages. Each stage drops off when its fuel is used up. As each stage leaves, the rocket gets lighter. It does not need so much power, so it does not use so much fuel. The engines can be lighter with each stage.

When the rocket is outside the planet's pull of gravity, only very small engines are needed. In outer space, the engines are used very little.

When the vehicle returns to the field of gravity, the engines act as brakes. They make the rocket fall more slowly.

Humans in Space

In May 1960 the Soviet Union sent a vehicle into space. There was a dummy astronaut in it. It did not come back to earth safely.

On August 19, 1960, the Soviet Union sent two dogs into space. Their names were Strelka and Belka. Their flight lasted one day. The were brought back by parachute.

The Russians made more test flights. Then, on April 12, 1961, the first human traveled to outer space. The Soviet spacecraft was named *Vostok*. The human was Flight Major Yuri Gagarin. His flight lasted 108 minutes. He went up about 187 miles (300 km). He orbited the earth once. For the first time, a human saw the earth from space. Then Gagarin landed safely back on earth.

Vostok 2

In August 1961 Major Gherman Tito orbited the earth in *Vostok 2*. His trip lasted 25 hours and 18 minutes. He orbited the earth seventeen times. Major Tito traveled more than 437,500 miles (700,000 km). He landed 450 miles (720 km) from the Russian city of Moscow.

In the United States, people who are pilots or crew members of a spacecraft are called *astronauts*. In Russia they are called *cosmonauts*. The first human in space was Russian cosmonaut Yuri Gagarin. He orbited the earth on April 12, 1961.

The first Soviet cosmonauts used the *Vostok* spacecraft. It contained a couch that protected the cosmonaut when the spacecraft speeded up. A heat shield kept *Vostok* from burning up when it came back to earth.

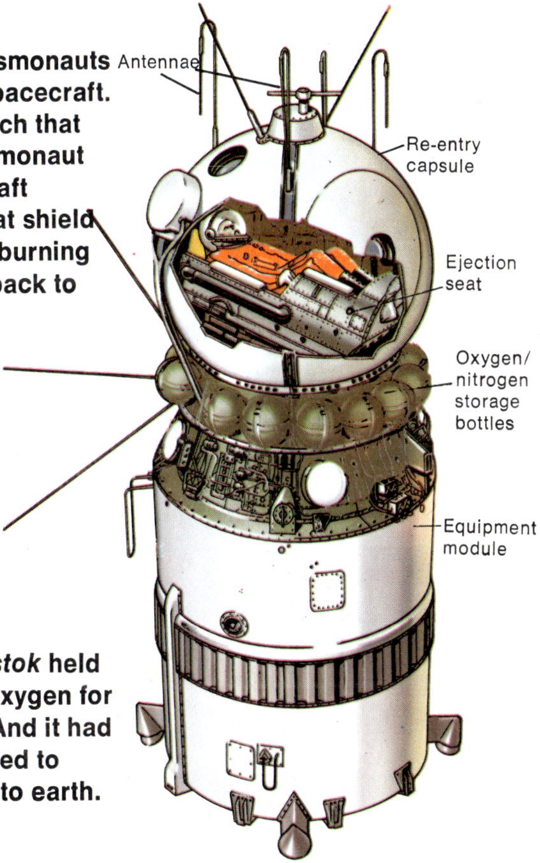

- Antennae
- Re-entry capsule
- Ejection seat
- Oxygen/nitrogen storage bottles
- Equipment module

The bottom of *Vostok* held food, water, and oxygen for the cosmonauts. And it had a rocket. It was fired to slow the fall back to earth.

On June 16, 1963, *Vostok 6* blasted off. Valentina Tereshkova was in *Vostok 6*. Tereshkova was the first woman in space. She made forty-eight orbits of the earth. She had been in training only a year. Before that, she worked in a cloth factory. Parachuting was her hobby.

United States Catches Up

Almost a month after Gagarin's flight, the U.S. sent up its first manned spacecraft. It was called *Friendship 7*. The astronaut was Alan B. Shephard. His flight lasted 15 minutes and 22 seconds. He did not orbit the earth.

On July 21, Captain Virgil Grissom made a space flight. Like Shepard, "Gus" Grissom did not go into orbit. His spacecraft was called *Liberty Bell*. There was trouble after *Liberty Bell* landed in the sea. Its escape hatch blew off. Grissom had to swim to safety. Two unmanned space flights were next. Both flights orbited the earth.

Then, for the first time, the U.S. put a human into orbit. He was Lieutenant Colonel John Glenn. He blasted off on February 21, 1962, in his *Mercury-Atlas* spacecraft. The spacecraft's steering was automatic. (It worked by itself.)

The *Mercury* capsule (spacecraft) used by the U.S. was shaped like a bell. This bell shape reduced the heat when *Mercury* returned to the atmosphere. (Atmosphere is the air around the earth.) The heat shield was made of layers of glass fiber and plastic. The retro-rocket pack was fired to slow down the landing.

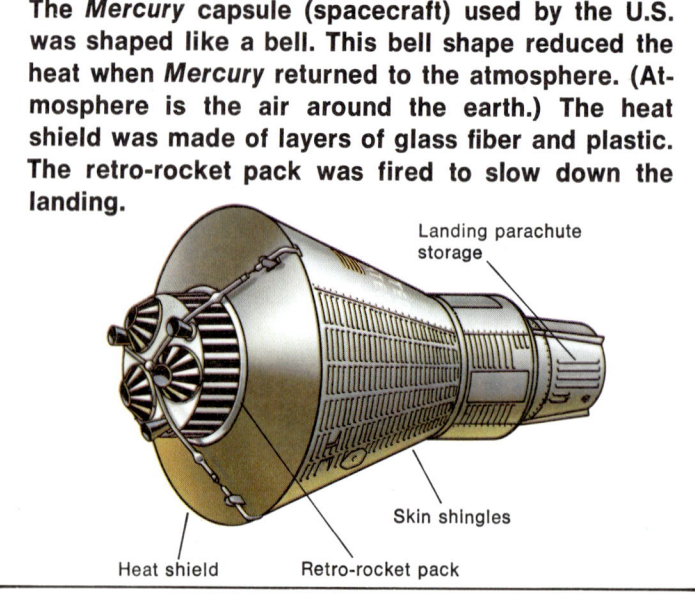

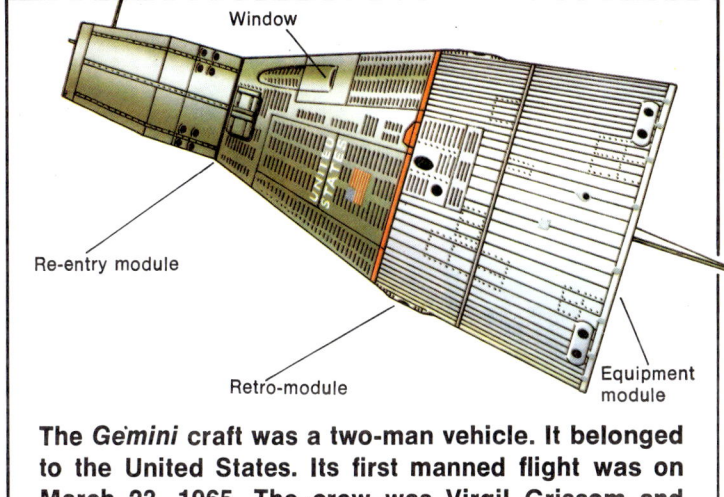

The Gemini craft was a two-man vehicle. It belonged to the United States. Its first manned flight was on March 23, 1965. The crew was Virgil Grissom and John Young.

But, when Glenn started his second orbit, something went wrong. Glenn had to steer the spacecraft himself.

More Problems

This was not Glenn's only problem. During his third orbit, there was a warning signal. The signal said the spacecraft's heat shield was loose. Was the signal right? If it was, the spacecraft could burn up while returning to earth. Ground control talked over the problem with Glenn.

Ground control decided not to fire the **retro-rocket.** They hoped it would hold the heat shield in place.

A retro-rocket is a small rocket at the front of a spacecraft. It is fired to slow down the spacecraft as it descends (comes down).

Glenn started down. He heard a thump. He saw fire through the window. He was afraid the heat shield was breaking. But he landed safely. Later, he found out the signal had been wrong. The heat shield was OK.

15

Walking in Space

On March 18, 1965, *Voskhod 2* lifted off from Russia. Two cosmonauts were on board. One of them was Lieutenant Colonel Alexei Leonov. He was the first human to walk in space. His walk took 23 minutes and 41 seconds. He performed some tasks in space. He showed that humans can work in space where there is no air.

A few weeks after Leonov's walk, Edward White stepped out of *Gemini 4*. He was the first U.S. astronaut to walk in space.

(Left) In 1965, Major Ed White walked in space. He was outside *Gemini 4* almost 125 miles (200 km) above the earth. Such walks helped astronauts prepare for landing on the moon.

(Below) The *Gemini* spacecraft was a two-man vehicle. It trained astronauts for the *Apollo* landings. While walking in space, astronauts got air through a cable.

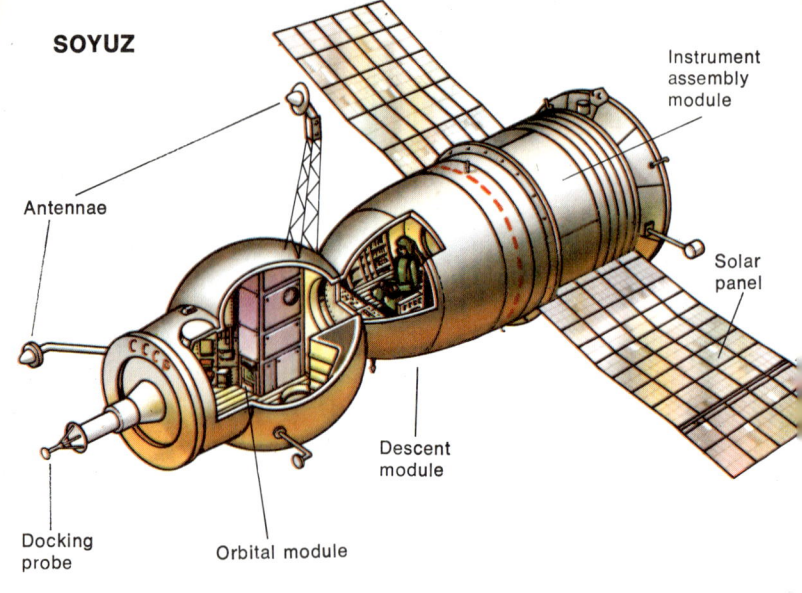

The First Space Station

Manned flights taught the United States and the Soviet Union a lot about space. By 1965 both countries were ready to begin **docking.** Docking is joining two spacecraft in outer space.

U.S. scientists knew this would help them land a person on the moon.

Russian scientists were more interested in learning how to build a big space station.

Meeting in Space
The Russians sent up their first *Soyuz* in April, 1967. It had a crew of three men. The *Soyuz* craft was in three parts, or **modules.**

The first module was where the men lived and worked. It was made for docking.

The second module housed the crew during take-off, descent, and landing.

The third module supplied the craft with power. Electric power came from batteries and solar panels. Solar panels changed sunlight into electric energy. The third module was made for communication, too. It had instruments to send and receive messages.

When it was landing, *Soyuz* got tangled up in the parachutes. It crashed to the ground. Cosmonaut Vladimir Komarov was killed.

The Russians took time out for safety tests. Then, on January 16, 1969, two crafts docked about 150 miles (240 km) above earth. This was the world's first space station.

One of the crafts was *Soyuz 4*. It came nose-to-nose with *Soyuz 5* and docked. The men from both ships performed some tasks in space. Then the two ships separated. They returned to earth.

A *Soyuz* spacecraft blasts off. The lower launch section was just about the same as *Vostok*'s.

Exploring the Moon

The United States wanted to land a manned craft on the moon before 1970. But first they had to learn more about the moon.

The United States sent out unmanned **space probes** to study the moon. (Probes send back information from outer space.) The probes were called *Lunar* (moon) *Orbiters*. Five were sent out in 1966 and 1967. They took photos of the moon's surface. These photos helped astronauts decide where to land.

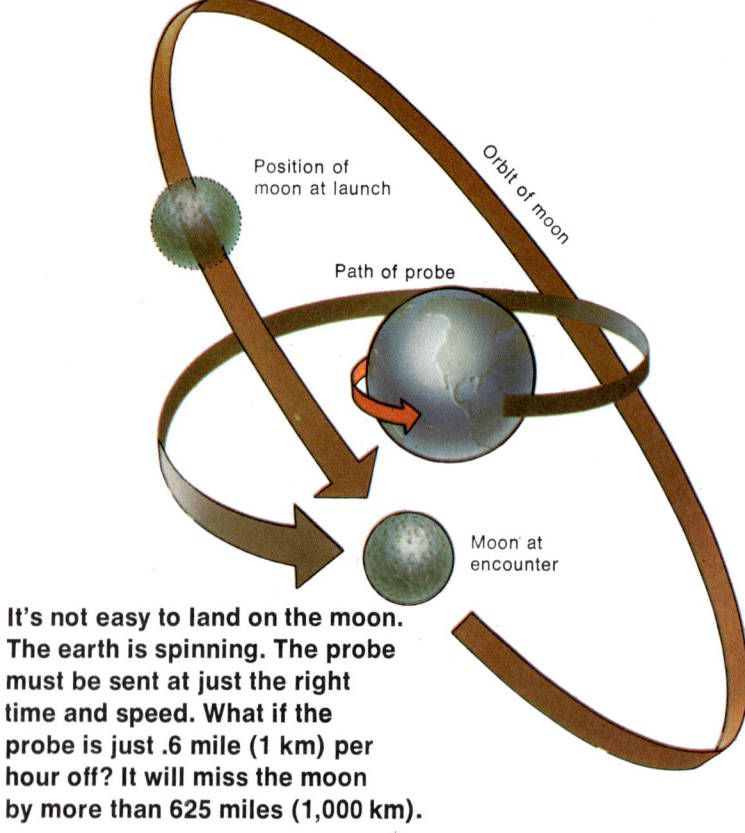

Position of moon at launch

Orbit of moon

Path of probe

Moon at encounter

It's not easy to land on the moon. The earth is spinning. The probe must be sent at just the right time and speed. What if the probe is just .6 mile (1 km) per hour off? It will miss the moon by more than 625 miles (1,000 km).

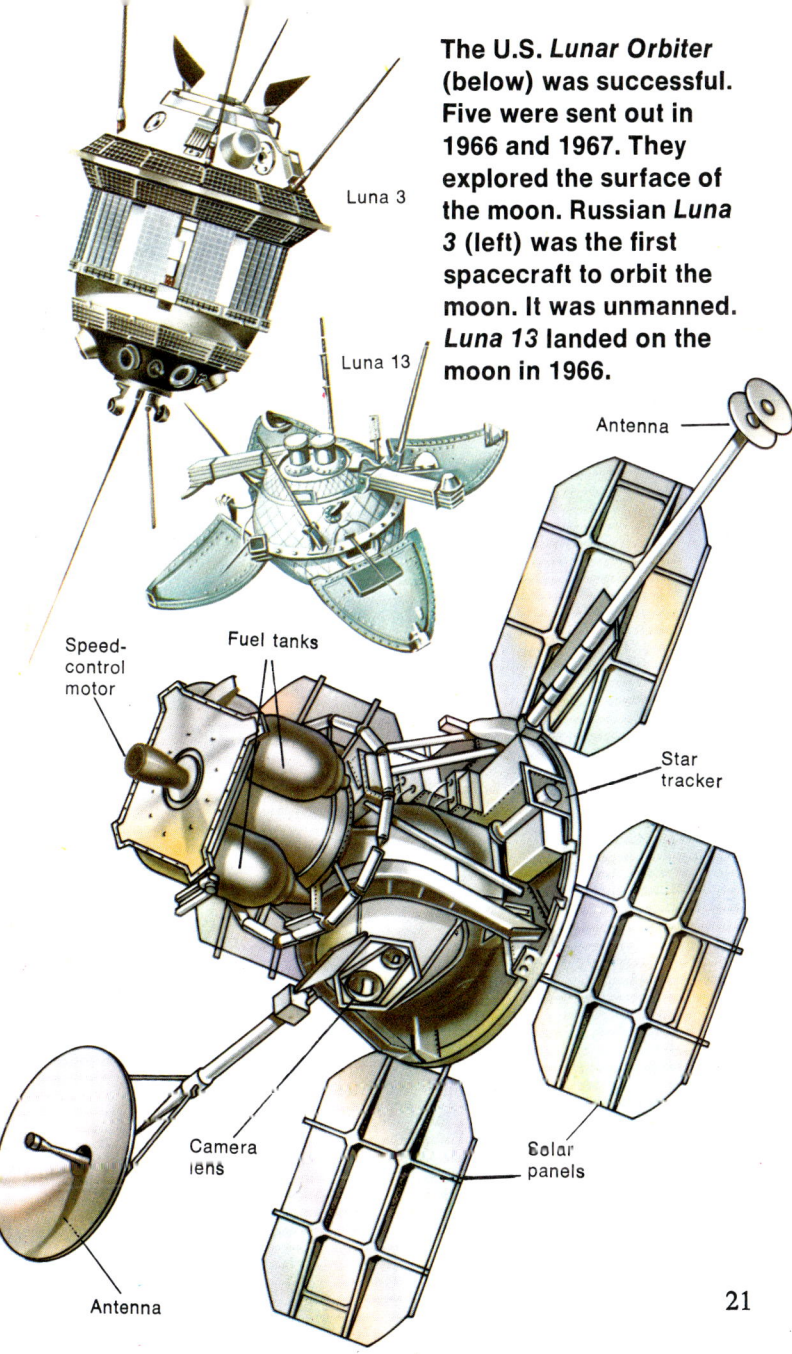

The U.S. *Lunar Orbiter* (below) was successful. Five were sent out in 1966 and 1967. They explored the surface of the moon. Russian *Luna 3* (left) was the first spacecraft to orbit the moon. It was unmanned. *Luna 13* landed on the moon in 1966.

Machines on the Moon

Lunar Orbiters and the Russian *Luna* series sent back many photos of the moon. The next step was to land equipment directly on the moon's surface.

In February 1966 *Luna 9* landed on the moon. It sent the first photos taken on lunar soil back to earth. In June the first of the U.S. *Surveyor* series landed. Just before Christmas, *Luna 13* landed. It measured the density (thickness) of the moon's soil. On April 17, 1967, *Surveyor 3* landed near the first *Surveyor*. It was the first probe to dig up soil from the moon.

On November 17, 1970, *Luna 17* left the first robot on the moon. It was called *Lunokhod 1* (below). Five men on earth controlled it. *Lunokhod* traveled across the moon for eleven months. It took photos and samples of the soil.

Surveyor

The U.S. *Surveyor* had three legs. It was almost 10 feet (300 cm) high. It could carry a lot of instruments and cameras. It had one main engine. This engine fired to control *Surveyor*'s ascent (climb upward). Three small jets were used for steering. A solar panel was on top of the *Surveyor.* This panel gave electric power for the instruments. Later *Surveyors* had long arms for digging the soil.

Seven *Surveyors* were launched between June 1966 and January 1968. *Surveyor 1* made the first controlled landing on the moon. But one of *Surveyor 2*'s engines failed on landing. *Surveyor 2* crashed. *Surveyor 4* lost radio contact with earth. It too was destroyed. Four other *Surveyors* landed safely. They completed their jobs.

A Close-up of the Moon

For many years people wondered what the surface of the moon was like. Some thought it was covered with rocks and huge craters (big holes). Others thought that dust covered the moon.

Then U.S. and Russian spacecraft began exploring the moon. We learned that both ideas about the moon's surface were partly right. Photos showed rocks of all shapes and sizes. And there was a thin layer of dust.

The land was of two kinds. There were places with old,

(Left) In 1966 *Lunar Orbiter 2* took this photo of the moon's crater, Copernicus. It is 56 miles (90 km) wide. Mountains in the center are nearly 219 miles (350 km) high.

Apollo 8 **astronauts took the photo on page 26. They took it on their 1968 trip around the moon. The photo shows the side of the moon that we can't see from the earth.**

light-colored rocks. These were "highlands." Places with newer, dark rocks were called "Maria" or "Seas." There were craters and mountain chains in both places.

Many scientists once thought some of the craters were the remains of old volcanoes. But scientists found nothing that showed volcanoes caused the craters. It seems that the craters were made by rocks crashing onto the surface. These rocks that fall from outer space are called **meteorites**. Some craters are more than 125 miles (200 km) across.

Many scientists thought that the moon was once part of the earth. They thought the moon broke away from the

25

earth millions of years ago. Studies of the moon show that this can't be true. Chemicals on the moon are different from those on earth. The moon may have come from somewhere else in the universe. The earth's gravity may have pulled it in.

This piece of rock came from the surface of the moon. The *Apollo* astronauts brought it back. Scientists think it is about 4.6 billion years old. It was made very early in the history of the moon. The surface of the moon is covered with rock chips and dust. There are larger pieces of rock all around. Tiny glass balls were also found. They were up to .02 inch (half a millimeter) across. Unlike earth's topsoil, there is nothing living on the surface of the moon.

Moon Rocks
Two main kinds of rocks were found on the moon. They are like the earth's *basalts* and *breccias.* Basalts are dark rocks with tiny crystals. Crystals are shiny minerals that look like ice. Breccias are lumps of rock chips, crystals, and soil. But the moon's rocks contain a lot of chromium, titanium, and zirconium. These materials are rare on earth. For example, moon rocks have 10 times more chromium than earth rocks.

The Apollo Program

In May 1961 President John F. Kennedy made an important announcement. He said the United States was trying to land a person on the moon by 1970. The *Apollo* spacecraft and the *Saturn 5* rocket were built. These would help the U.S. achieve its goal.

The plan for the *Apollo* moon flights. This plan is called *lunar orbiter rendezvous*. Rendezvous means bringing two spacecraft together. The only *Apollo* flight that did not follow this plan was *Apollo 13*. It failed.

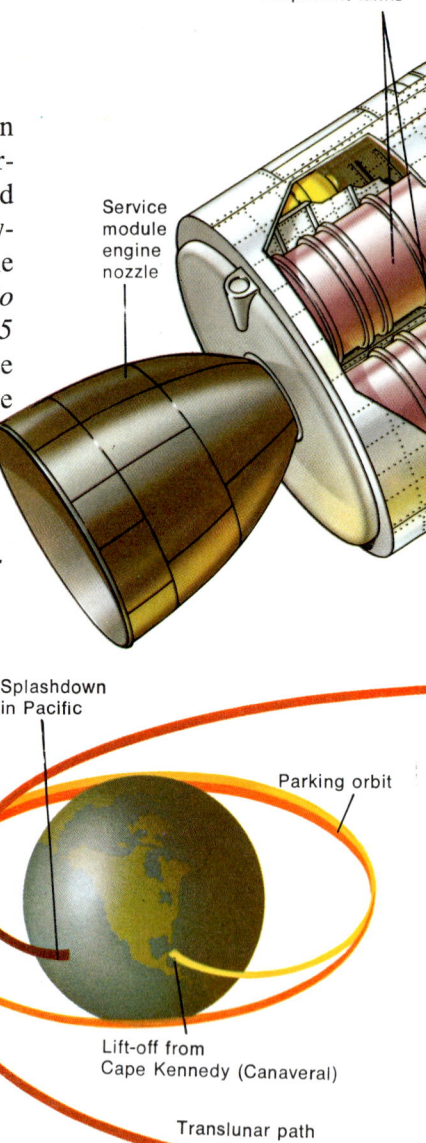

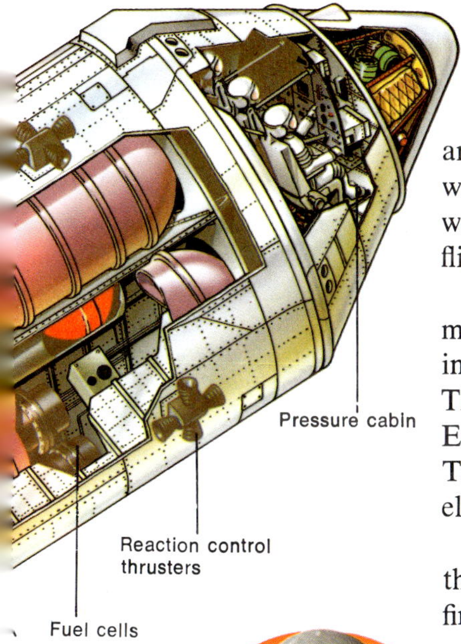

Command module

On February 26, 1966, an *Apollo* command module was launched. Then there were five more unmanned flights.

Next, three astronauts made the first manned flight in the *Apollo*. It was in 1968. The astronauts were Schirra, Eisele, and Cunningham. They orbited the earth for eleven days.

About two months later, three astronauts made the first manned flight to the

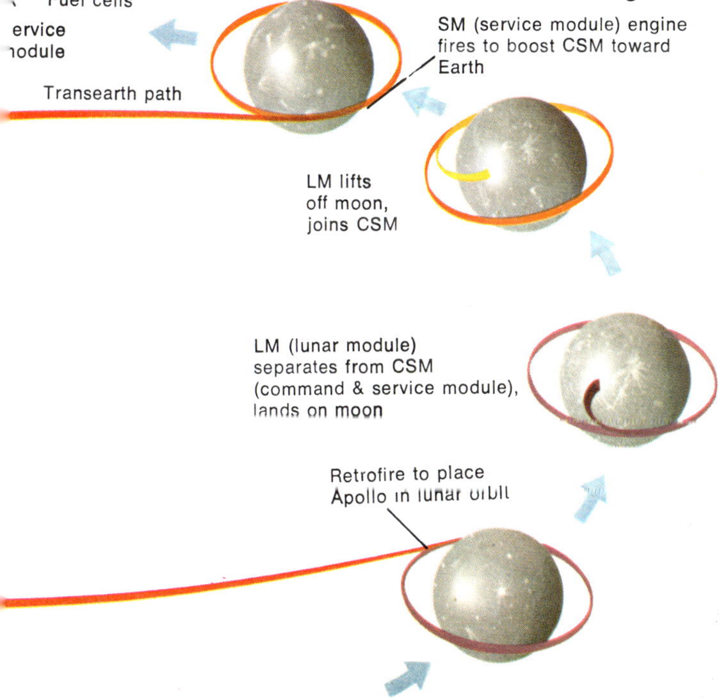

Pressure cabin

Reaction control thrusters

Fuel cells

Service module

Transearth path

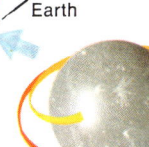

SM (service module) engine fires to boost CSM toward Earth

LM lifts off moon, joins CSM

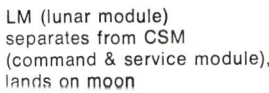

LM (lunar module) separates from CSM (command & service module), lands on moon

Retrofire to place Apollo in lunar orbit

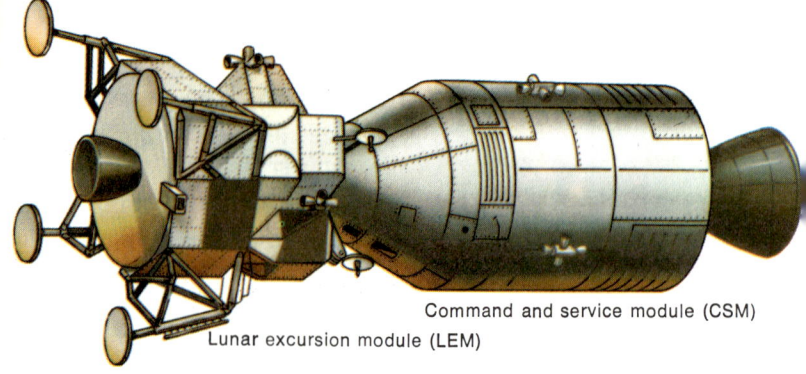

Command and service module (CSM)
Lunar excursion module (LEM)

moon. These astronauts were Borman, Lovell, and Anders. They made ten orbits before they came back safely to earth.

Stafford, Young, and Cernan were the next astronauts to orbit the moon. They made a test descent (trip down) toward the moon's surface. The landing module came to within 9 miles (14 km) of the moon's surface. Then the astronauts went back to dock with the command module.

Man on the Moon

It was July 20, 1969. Neil Armstrong and "Buzz" Aldrin, Jr. were the *Apollo 11* astronauts. The lunar module had landed on the moon.

The astronauts were getting ready to leave the module. It took hours. At last everything was done.

Armstrong stepped onto the surface of the moon. Millions of people all over the world heard his words: "That's one small step for a man; one giant leap for mankind."

Safely Home

Aldrin went out, too. The gravity was low. The men practiced moving about. They set up some instruments. They took samples of rock and soil.

About 3½ hours later they docked with the command module. They got back to earth on July 24.

(Left) The *Apollo* spacecraft had three parts. At one end was the lunar module with four small legs. It was joined to the command module (shaped like a cone). The long round part was the service module and engine.

(Below) The lunar module was made of aluminum. It had two main parts. The lower part had four legs. It held the descent engine. This was left behind when the top part blasted off.

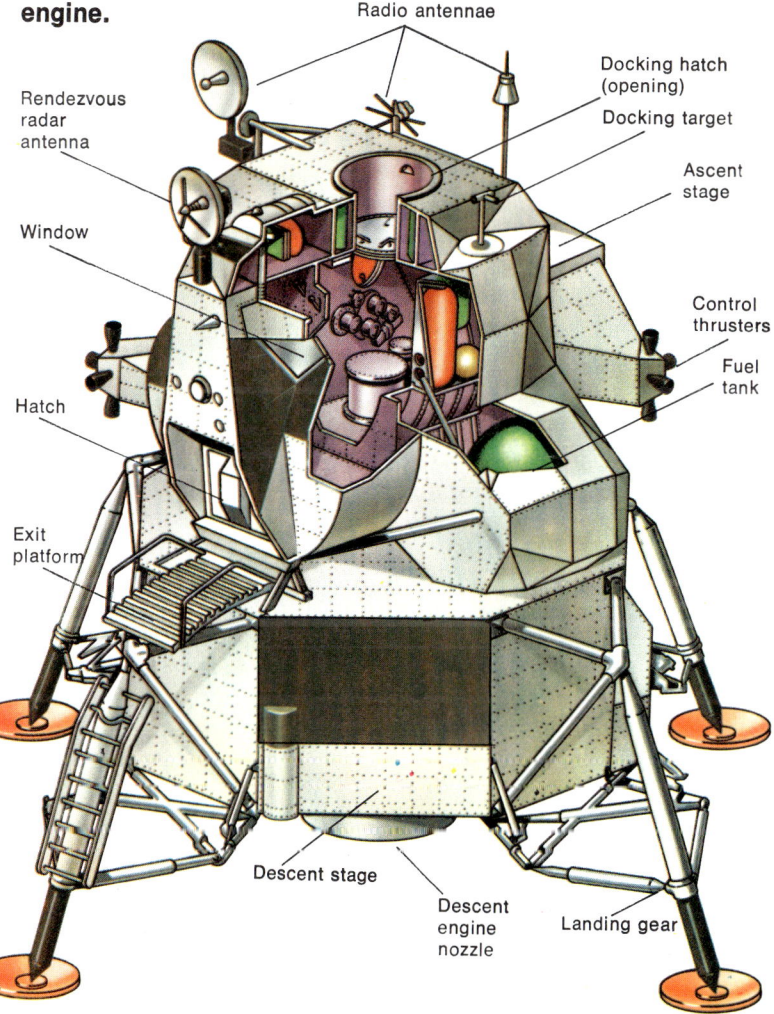

To the Moon and Back

(1) A trip to the moon starts with launching the *Saturn* rocket to put the *Apollo* spacecraft in orbit. (2) The command, service, and landing modules pull out of their cover. They head for the moon. (3) They begin orbiting the moon. The astronauts get into the landing module. The landing module pulls away. (4) It descends to the surface of the

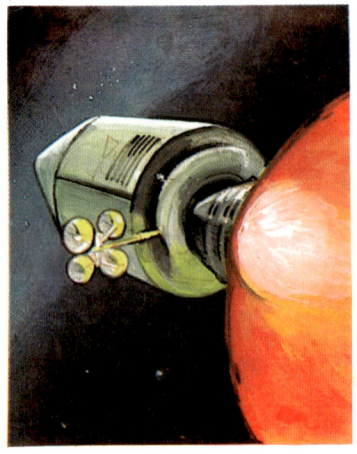

3

4

moon. The command and service modules stay in orbit. (5) The astronauts return by blasting off in the top part of the landing module. (6) They join the command module again. They head back to earth in the command module. (7) They get near the earth's atmosphere. The service module drops. The command module starts down. (8) Parachutes help it go slower for splashdown.

7

8

33

Space Suits

The higher you go above the earth, the less air there is. In outer space there is no air at all. And outer space can be very hot or very cold. So astronauts need space suits to survive.

A space suit is like a small spacecraft. It can do a lot of things. Space suits give astronauts oxygen to breathe. This oxygen may come from tanks in the space suits. Or it may come through a hose attached to the spacecraft. Space suits also remove the carbon dioxide that the astronauts breathe out. And the inside of the space suit is at just the right temperature. Astronauts are protected from extreme heat or cold.

Space suits can get rid of body wastes. They can check to see if the astronaut is healthy. They can even send and receive radio messages.

Most space suits have several layers. The inside layer keeps the astronaut's body at the right temperature. The second layer stops the fast change in speed from harming the astronaut's body. The next layer keeps the space suit from being torn or damaged.

When astronauts are in a spacecraft, they usually don't need space suits. The spacecraft is made to take care of their needs. Space suits are big and lumpy. Astronauts are glad to do without them. But sometimes astronauts could not survive unless they wore them. Astronauts need space suits when the spacecraft takes off and lands. They also need space suits in emergencies.

Keeping the Right Pressure

A space suit keeps an astronaut's blood pressure normal. Blood pressure is the force of the blood against the inside

(Right) Here is Edwin "Buzz" Aldrin on the moon. Neil Armstrong took this photo after *Apollo 11* landed.

Edwin Aldrin climbs down to join the first man on the moon. (Right) Apollo space suits protected the astronauts from the freezing cold and boiling heat of space. The space suits gave the astronauts air. They could even send and receive radio messages. And the suits made sure the fast change in speed did not hurt the astronauts.

of the blood vessels. What if you were about 8,900 yards (8,000 m) above the earth? Without a space suit, your blood would boil! The oxygen in your blood would turn to gas. This gas would take up more room than liquid blood. Most adults have about 8 pints of blood in their bodies. If the blood became gas, it would take up as much room as 24 pints. Your body would explode!

Space suits protect astronauts from sudden changes in speed. What happens when you are in a car that starts very fast? You are pushed back in your seat. All of a sudden it feels like you weigh much more. When a rocket is launched, the speed increases very quickly. It goes up to 25,000 miles (40,000 km) per hour. Without a space suit, astronauts could not stand the sudden gain in weight. They would die.

Protective helmet

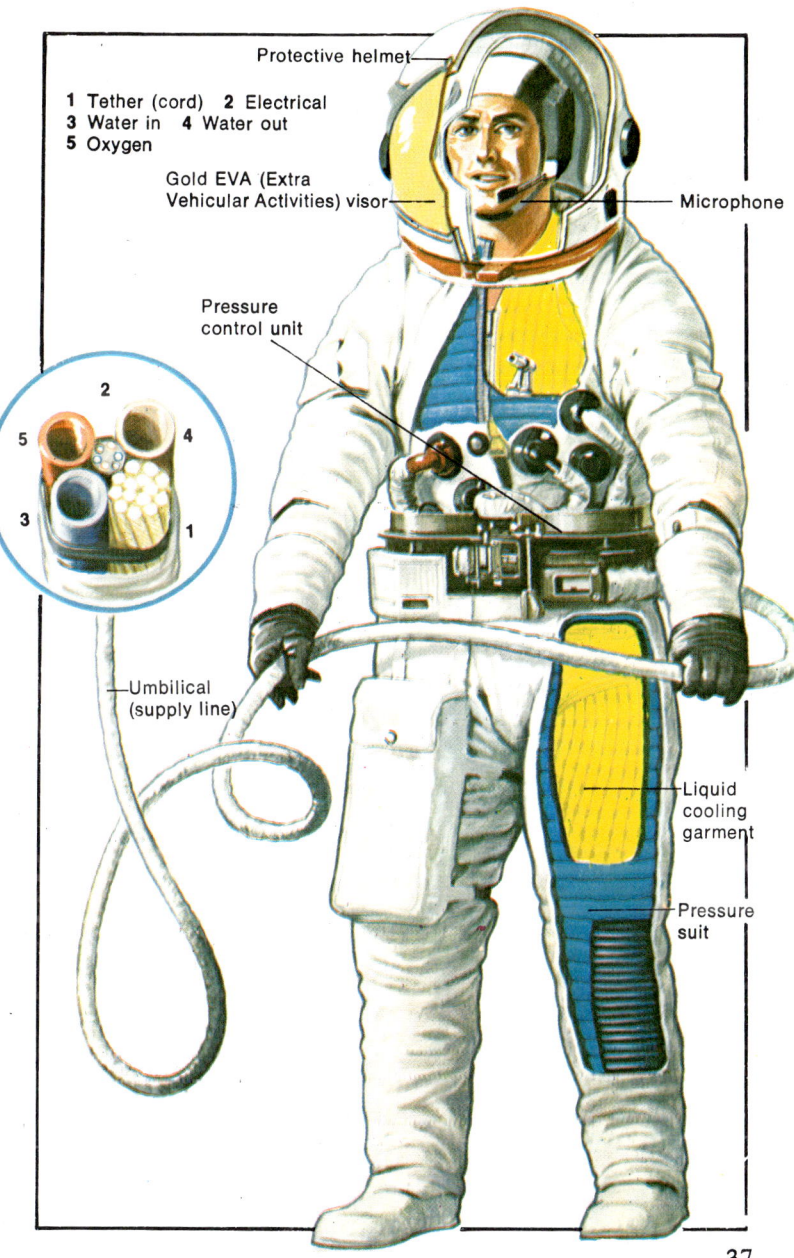

1 Tether (cord) **2** Electrical
3 Water in **4** Water out
5 Oxygen

Gold EVA (Extra Vehicular Activities) visor

Microphone

Pressure control unit

Umbilical (supply line)

Liquid cooling garment

Pressure suit

37

Apollo 12

In November 1969 *Apollo 12* made the second moon landing. Conrad, Gordon, and Bean were its crew.

Lightning had struck the *Saturn* rocket on its launchpad. So the flight was almost called off. But finally *Apollo 12* was launched.

Conrad and Bean landed near the old *Surveyor 3* probe. They spent more than 31 hours on the moon.

Lunar module from *Apollo 12* rests on moon. Gold foil protects the module from the sun's heat.

Apollo 13

Apollo 13 was the third moon-landing flight. It's astronauts were to study the surface of the moon very carefully.

But the flight was in trouble from the start. Astronauts Lovell, Mattingly, and Haise were exposed to German measles. The flight was almost called off. Finally doctors said Lovell and Haise were all right. But, another astronaut, Jack Swigert, had to take Mattingly's place.

Then the launch went wrong. One engine in the second stage shut down too soon. But Mission Control was able to get more power from the other four engines. (Mission Control is the crew that controls spaceflights. It is on the ground.) The last stage pulled away without any problems. It headed for the moon.

Fifty-six hours went by. Everything seemed fine. Then there was a bang! The command module's electrical and life-supply systems failed.

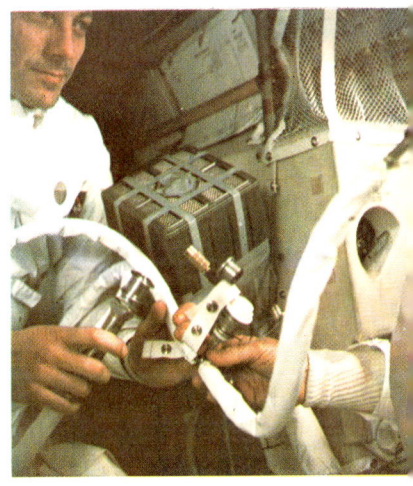

The *Apollo 13* air conditioning failed. Carbon dioxide was building up. The astronauts made this air conditioner by themselves.

An oxygen tank in the service module had exploded. The trip to the moon was called off. Now the astronauts had to get back to earth.

They used the lunar module as a lifeboat during the trip back. The air conditioning could not take out all the poisonous carbon dioxide. The astronauts had to make another air conditioner.

At last *Apollo 13* returned safely to earth. The splashdown was successful.

(Above) Emblem of *Apollo 17*

(Right) This photo shows *Apollo 17*'s command and service module. It was taken from the lunar lander.

The Last Apollo

Apollo 17 was launched in December 1972. It was the last moon landing. The crew was Commander Eugene Cernan, Ronald Evans, and Dr. Harrison Schmitt. Dr. Schmitt was a geologist. He had studied the crust of the earth.

There were problems when *Apollo* tried to take off. Some of the equipment was not working properly. It seemed as though the flight would be called off. But 2½ hours later, the *Saturn 5* rocket lifted off safely, launching *Apollo 17* into space.

On December 10, *Apollo 17* began orbiting the moon. It orbited eleven times. Then the lunar module pulled away from the command module. Cernan and Schmitt were in the lunar module. It landed on the moon's surface.

In December 1972 astronaut Eugene Cernan traveled over the surface of the moon. Cernan traveled in the lunar rover. It saved time and energy. It was left on the moon.

Four hours later, the astronauts went out of the module. They set up instruments. One instrument recorded the amount of heat coming from inside the moon.

The two astronauts drove around in their moon rover. The moon rover had four wheels. It was powered by two 36-volt batteries. Each wheel had a motor. The moon rover could be folded up for the flight out. The astronauts could go far in the moon rover. And they did not have to use much energy, water, or oxygen.

The moon rover could carry two times as much as it weighed. So the astronauts could take a lot of equipment with them.

Mission Control watched the two men while they explored. The crew watched through a remote (faraway) control TV camera.

Apollo 17 broke many records. It was on the moon's surface longer than any other spacecraft. It stayed 74 hours, 58 minutes, and 38 seconds. The astronauts were out of the lunar module longer than any other astronauts. They were out for 22 hours, 5 minutes, and 6 seconds. They made the longest trip on the moon's surface. It lasted 7 hours, 37 minutes, and 21 seconds. *Apollo 17* brought back about 220 pounds (100 kg) of moon rock samples. This was the most moon rocks ever brought back.

The Happy Ending
The moon rover was left on the moon. A TV camera had been set up on it. This camera showed the astronauts' lift off. They went back to the command module.

The astronauts spent two days making maps. These maps were of the far side of the moon.

Then *Apollo 17* came home. The entire trip took 12½ days. It ended safely with splashdown in the Pacific Ocean.

(Left) *Apollo 17*'s Dr. Harrison Schmitt works by a huge moon rock. The moon rover is parked in front.

Photo taken on moon. U.S. flag and faraway earth.

43

Salyut

The Americans and the Russians worked hard to learn about spaceflights. But they put their knowledge to different uses. The United States wanted to land on the moon. So they began the *Apollo* flights. The Russians wanted to put space stations in orbit around earth. So they began the Salyut program.

Space Stations
Salyut is a Russian space station. It was put in orbit in April 1971. Space stations are very large. Astronauts can stay in a space station for many days. This gives them time to learn a lot about space.

There are several *Salyut* models. Most are long, round parts (cylinders) joined together.

The *Soyuz* spacecraft was used as a shuttle. It went back and forth between the earth and *Salyut*. It took men and supplies to *Salyut*.

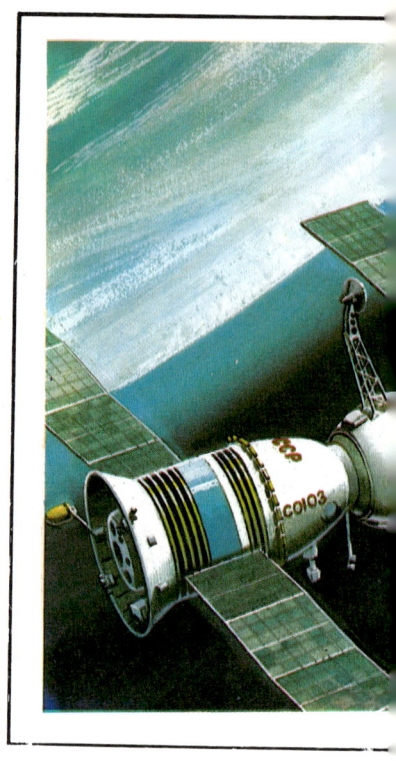

When *Soyuz* was docked, it became another part of the space station. When docked, the entire station was 22 yards (20 m) long.

The crew lived and worked in the main part of *Salyut*. Fuel tanks were built on the end of *Salyut*.

The crew of *Soyuz 11* was the first to enter *Salyut*. They checked *Salyut* and did some experiments. They grew some plants to find out how well they would survive in space.

After twenty-three days the *Soyuz 11* crew started back to earth. But a pressure seal failed. The cosmonauts died.

For two years the Soviet Union took time out to improve safety. Then *Salyut 2* was launched. Other *Salyut* models came later. In 1979

Russian space station. Cosmonauts stayed in this station more than five months. At the left is the *Soyuz*. It went back and forth between earth and the station. It carried men and supplies.

two *Salyut* cosmonauts spent 175 days in orbit.

45

Telescope and solar observation unit

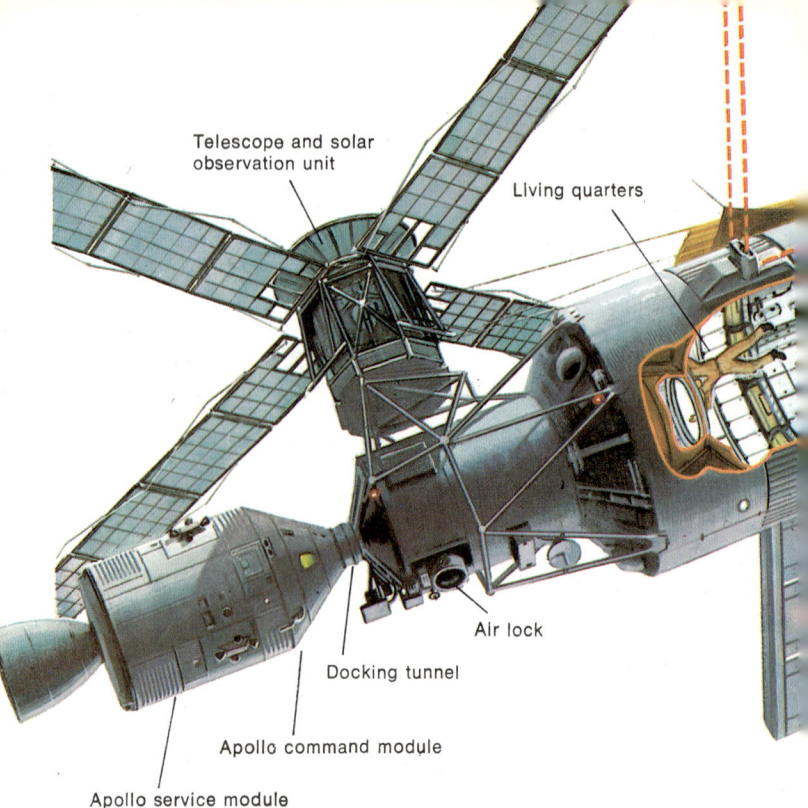

Living quarters

Air lock

Docking tunnel

Apollo command module

Apollo service module

Skylab

Skylab was the United States' first space station. It cost a lot of money and took many years to build. *Skylab* was launched in May 1973. About a minute after lift-off, a heat shield was torn away. This shield was to protect *Skylab*'s workshop. At the same time, a solar power panel was carried away. Another panel was damaged.

An *Apollo* spacecraft was launched to save *Skylab*. It carried special tools and sun shields. The sun shields would keep the inside of *Skylab* from getting too hot. *Apollo* reached *Skylab*. The crew checked to see what was wrong.

It was hard to dock. But *Apollo* made it. Then the crew

Orbital workshop

Solar panels

In the beginning *Skylab* had many problems. But repairs were made. Then *Skylab* performed well. The crew did many experiments. Maybe the most important experiments were those with the *Apollo* telescope. For the first time people studied the sun, moon, stars, and planets from a platform in space.

went into *Skylab*. The astronauts wore face masks. The masks protected the men from any dangerous gases.

The men put up a heat shield that looked like an umbrella. It protected them from heat. Then they fixed the solar panel. On June 22 the astronauts undocked. They returned to earth.

A second crew was sent to *Skylab* a month later. They took mice, insects, and fish with them. Scientists wanted to find out if animals could live in space. The astronauts did some experiments with the animals. They also made some repairs on *Skylab*.

On August 7 two of the crew put on their space suits. They went out to put a second heat shield over the first one. The heat shield made *Skylab* cooler, about 75°F (24°C). Now it was easier for the crew to do experiments.

The wiring in the command module became damaged. This killed the mice and some insects. But the spiders went on spinning their webs. The fish learned to swim in their new home.

The crew came back to earth on September 25. They had gone 25 million miles (40 million km). The astronauts returned with 17,000 photos of the earth, 75,000 photos of the sun, and more than 62 miles (100 km) of tapes with what they had learned.

(Right) Kerwin and Conrad of *Skylab 1.* There is no pull of gravity in space. So objects don't have weight. They float in the air.

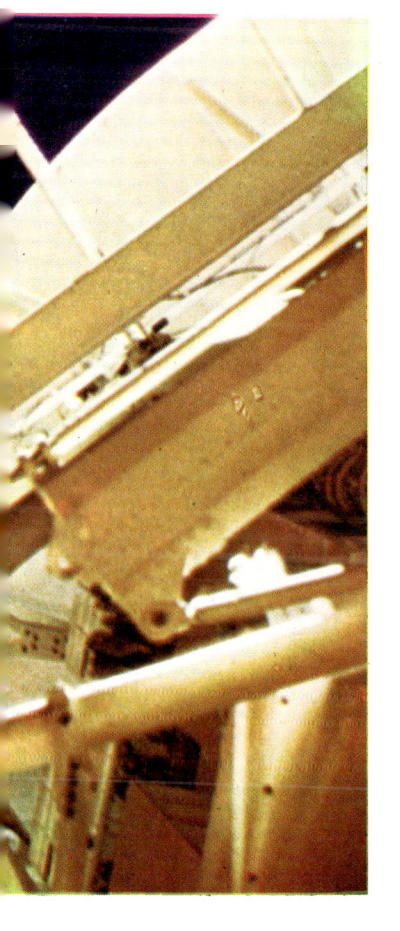

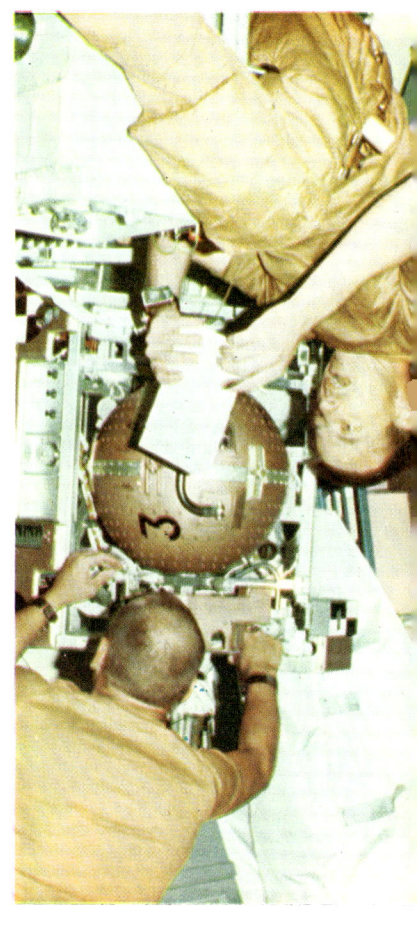

(Left) Owen Garriott was one of *Skylab*'s crew. Here, he loads film into the telescope.

Saturn was the third and last spacecraft to visit *Skylab*. The crew docked with *Skylab* on the third try. They planned to stay on board eighty-four days.

For the first time, astronauts had a chance to study a **comet** in space. Comets are bright objects that orbit the sun. Kohoutek is a giant comet. It orbits the sun once every 80,000 years. *Skylab*'s astronauts were able to watch it move behind the sun.

The crew did a lot of other important work on *Skylab*. They orbited the earth 1,213 times. They spent 22 hours and 19 minutes outside the space station. Then, in February, they splashed down in the Pacific Ocean.

After this visit *Skylab* was shut down. Scientists thought it would stay in space about six to ten more years. But it came back into the atmosphere early. It broke into pieces in July 1979. These pieces broke into smaller pieces. **They dropped to the surface of the earth.**

(Above) Three teams of astronauts visited *Skylab* in 1973 and 1974. They splashed down in the Pacific. Here, frogmen get a capsule ready to be lifted onto a ship.

(Left) Taking a shower in space. Special equipment keeps the water from floating all over!

(Right) Checkups are needed in space. Here, Dr. Kerwin takes a look at Conrad's mouth.

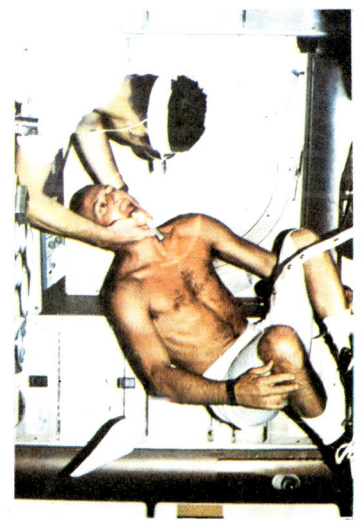

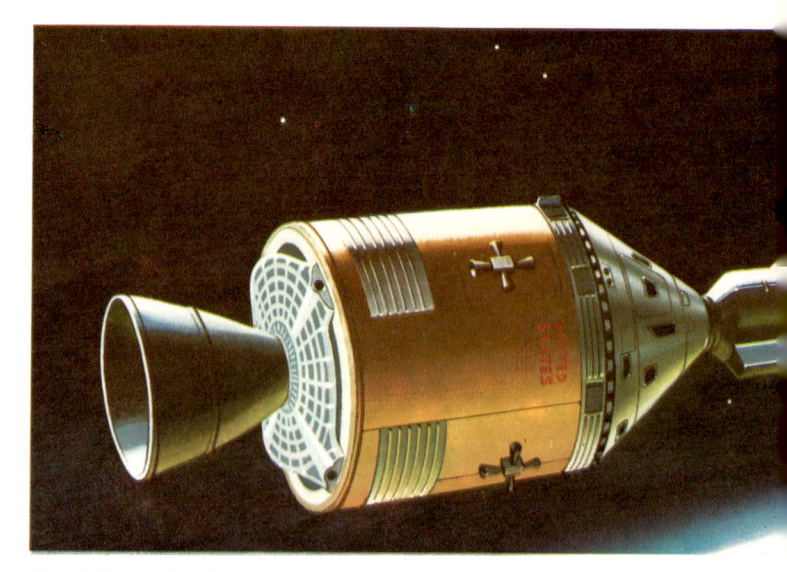

A Handshake in Space

On May 24, 1972, the United States and the Soviet Union signed an agreement. Both countries agreed to use space for peaceful purposes.

They planned a spaceflight together. This flight was called the Apollo/Soyuz project. It was to take place on July 15, 1975.

Apollo would dock with *Soyuz*. Both countries had to work together. Special parts had to be made.

The Great Day

At last the great day came. The *Soyuz* spacecraft was launched about 7½ hours before *Apollo*. *Soyuz* orbited and waited for *Apollo*. *Apollo* was carrying a special docking module. About 50 hours later, *Apollo* was nose-to-nose with *Soyuz*. The two craft became one.

The seals were taken off the spacecraft's hatches (openings). A tunnel connected the two craft. Leonov of the

In 1975 a U.S. *Apollo* and a Soviet Union *Soyuz* joined in space.

Soviet Union and Stafford of the United States shook hands. Later they ate together. For two days the two crews worked side by side. They studied and did experiments. Then the two craft separated.

Forty-three hours after leaving *Apollo*, *Soyuz* returned to earth. *Apollo* stayed in orbit for about 3½ days. Then it splashed down near Hawaii in the Pacific.

A Promise to Help

The United States and the Soviet Union decided to help each other in space. Spacecraft would be made so that they could contact each other in an emergency.

Looking Down on Earth

Astronauts can see many things on earth while in orbit. Gordon Cooper orbited the earth in *Gemini 5*. He saw the wakes of ships in the Atlantic Ocean. (The wake is the trail left behind a moving ship). And once he saw the shape of an airliner. It was landing at El Paso International Airport.

This photo of the earth was taken from *Apollo 11* in July 1969. Pictures like this show what the weather will be, and they warn of dangers such as hurricanes. They show where warm air and cold air will meet and form rain.

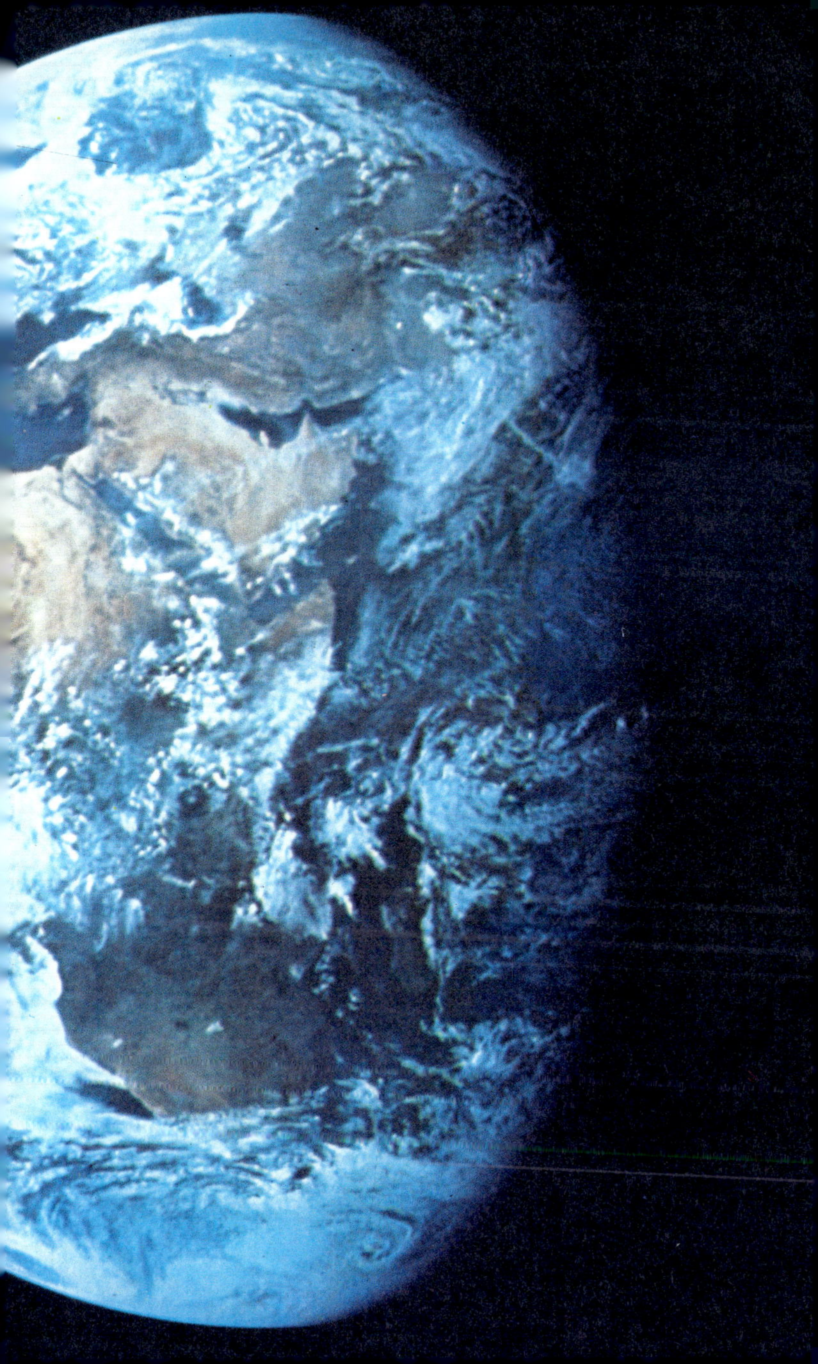

(Above) *Gemini 11* astronauts took this photo. It shows the Sinai Peninsula and the Gulf of Suez.

(Below) An unmanned satellite took this photo. The satellite was about 570 miles (about 912 km) above the earth. Satellite photos help scientists learn more about the earth's crust.

Man-made satellites and manned flights have helped us learn more about earth. Scientists have learned many new, special ways to take photos. Scientists study these satellite photos to learn about the earth's crust and the oceans. The photos also help us predict what the weather will be. They warn us of storms.

(Left) This photo was taken from *Gemini 7.* It shows part of the Sahara Desert. Space photos help mapmakers draw accurate maps.

57

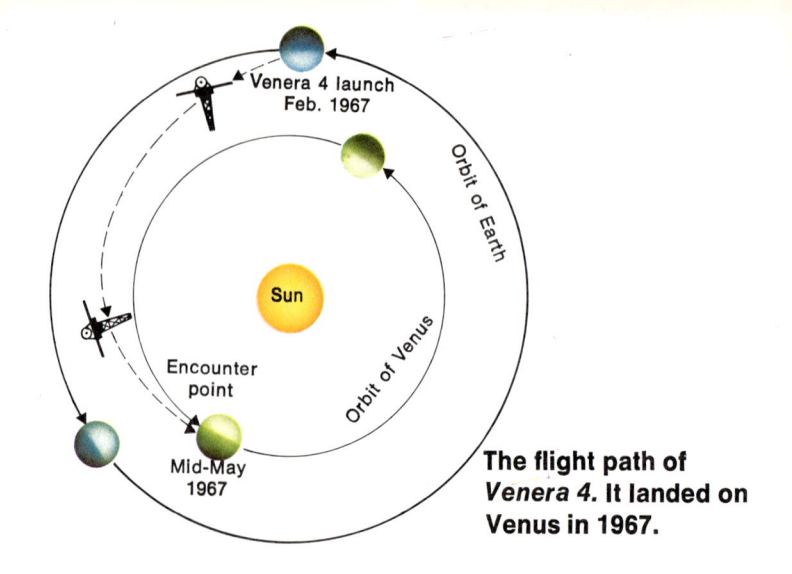

The flight path of *Venera 4*. It landed on Venus in 1967.

Probing the Planets

A **probe** is an unmanned spacecraft. It sends information from space back to earth. A planetary probe sends back information about a planet. The probe may fly past the planet. It may orbit or land on the planet.

The Soviet Union studied the planet Venus. This was difficult because Venus is very hot. Russia's *Venera 4* was the first probe to send back information about Venus. *Venera 4* was followed by six more *Veneras*.

Close-up of the surface of Venus. *Venera 9* took the photo.

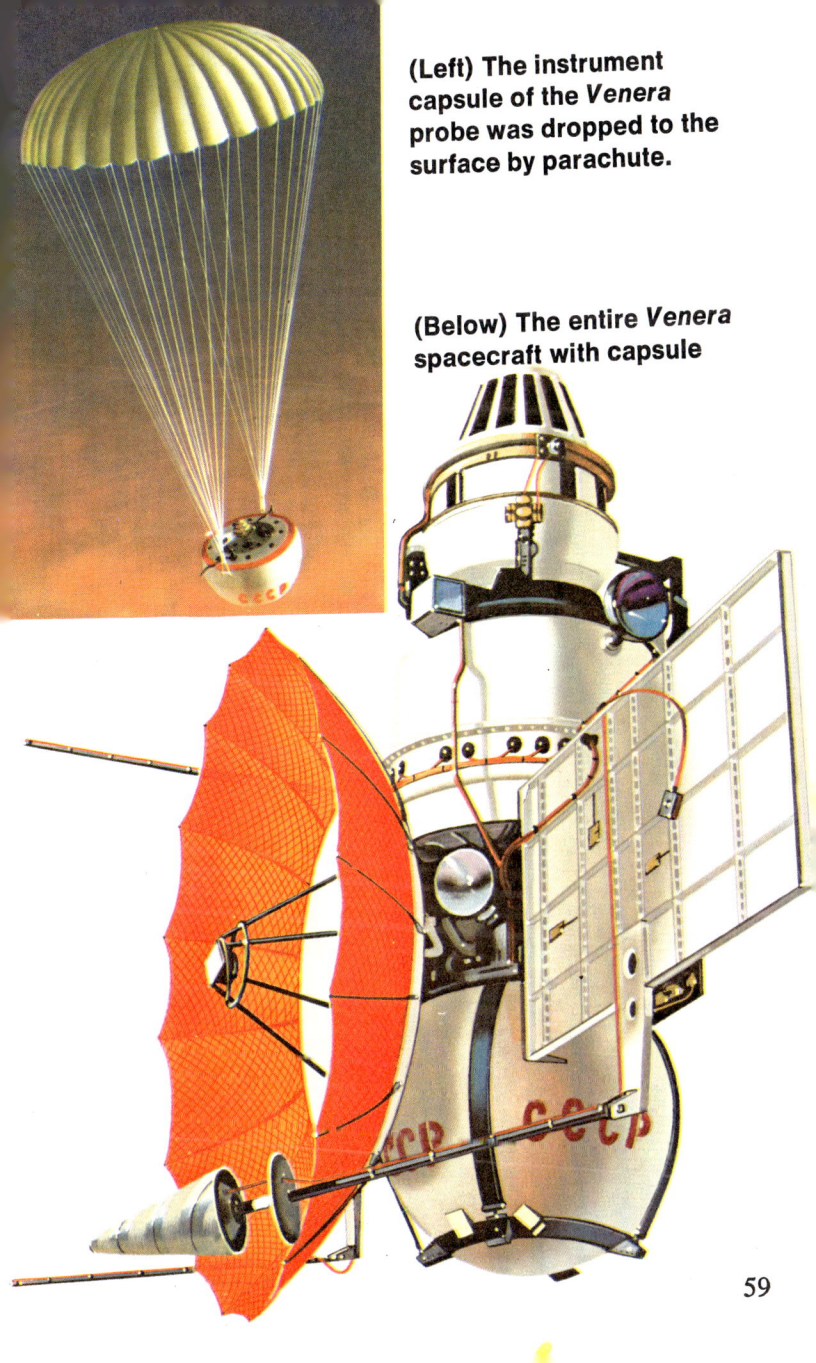

(Left) The instrument capsule of the *Venera* probe was dropped to the surface by parachute.

(Below) The entire *Venera* spacecraft with capsule

Probes to Jupiter

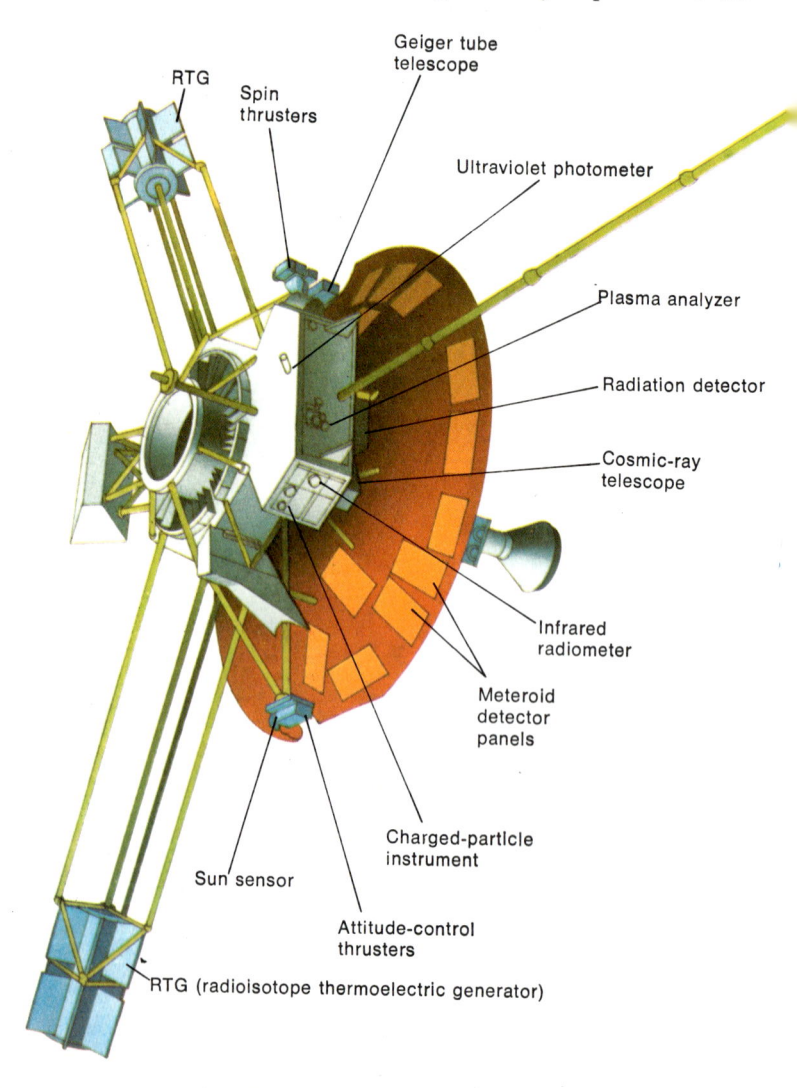

Pioneer 10 was the first probe to explore Jupiter. The *Pioneer* was launched by the United States in March 1972. It was 81,250 miles (130,000 km) from Jupiter when it flew past. In 1974, *Pioneer 11* passed by Jupiter at a dis-

tance of 25,000 miles (40,000 km). The *Pioneer* was going more than 106,250 miles (170,000 km) per hour. It was the fastest object humans had ever made. Next, *Voyagers 1* and *2* passed by Jupiter on the way to Saturn.

—— Magnetometer

(Left) The *Pioneer* probe studied Jupiter. Jupiter is known as the giant planet. *Pioneer* studies show a liquid surface.

(Below) *Pioneer 11* took this photo in 1974. It shows Jupiter's Red Spot. The Spot may be a whirlwind bigger than earth.

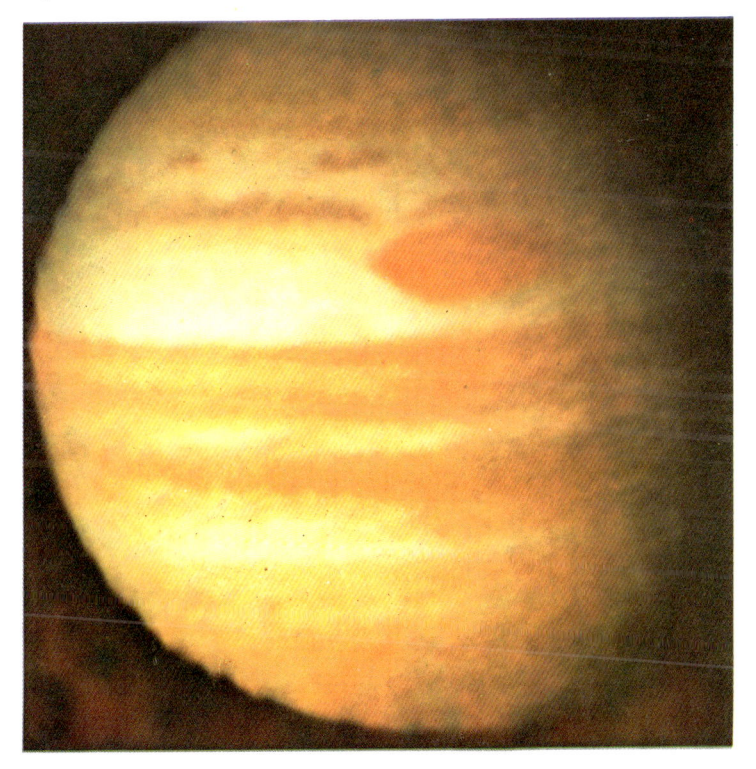

Visits to Mars

What is Mars really like? People asked this for many years. They made up stories. Finally, U.S. and Russian probes gave us some facts. They showed that Mars has no water. Its atmosphere is thin. There is no sign of life.

The U.S. probes were called *Mariner* and *Viking*. Soviet Union probes were called *Mars*.

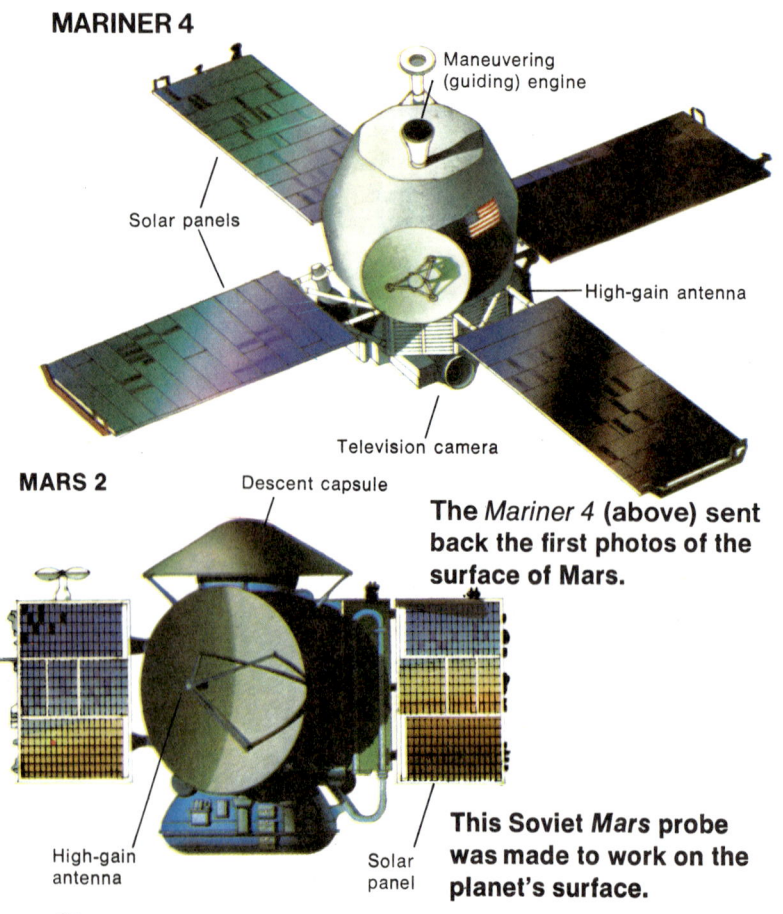

MARINER 4

- Maneuvering (guiding) engine
- Solar panels
- High-gain antenna
- Television camera

MARS 2

- Descent capsule
- High-gain antenna
- Solar panel

The *Mariner 4* (above) sent back the first photos of the surface of Mars.

This Soviet *Mars* probe was made to work on the planet's surface.

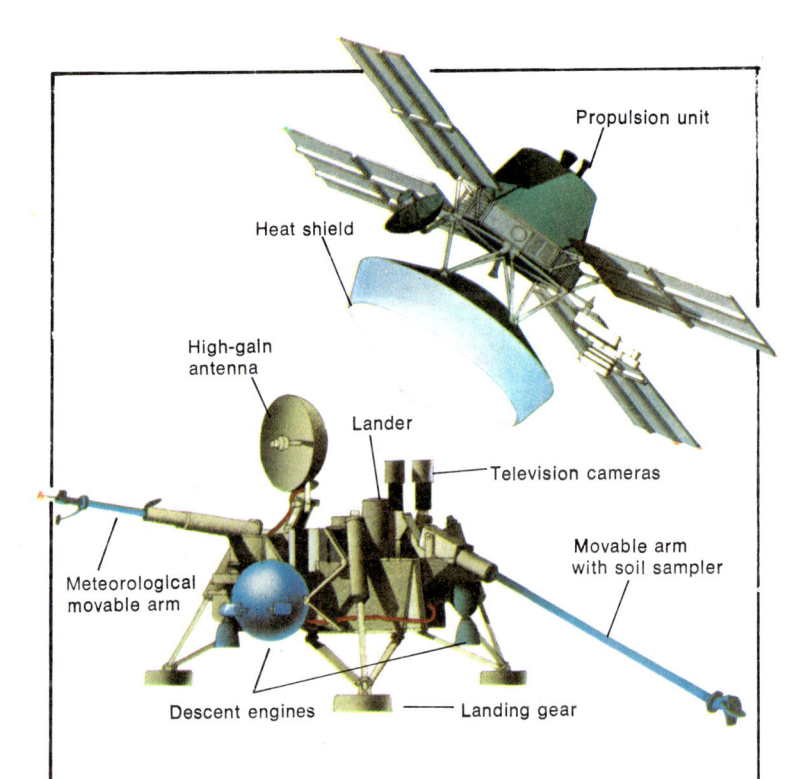

The Vikings

Viking 1 (above) landed on Mars on July 20, 1976. It sent back the first photos of the surface of Mars. And it tested the soil.

Viking 2 was launched on September 9. It touched down nearly a year later. It landed in a different spot than *Viking 1*. It was warmer there, and the winds were milder. *Viking* had an arm that could examine samples of the soil. No signs of life were found.

Hot Mercury

Mercury is the planet nearest the sun. It's surface is about 752°F (400°C). It's so hot nothing can live there. Mercury is almost 62 million miles (100 million km) from the earth. It is so tiny and far away that it's hard to see. Scientists did not know much about it. Then, on November 3, 1973, *Mariner 10* was launched to explore the planet.

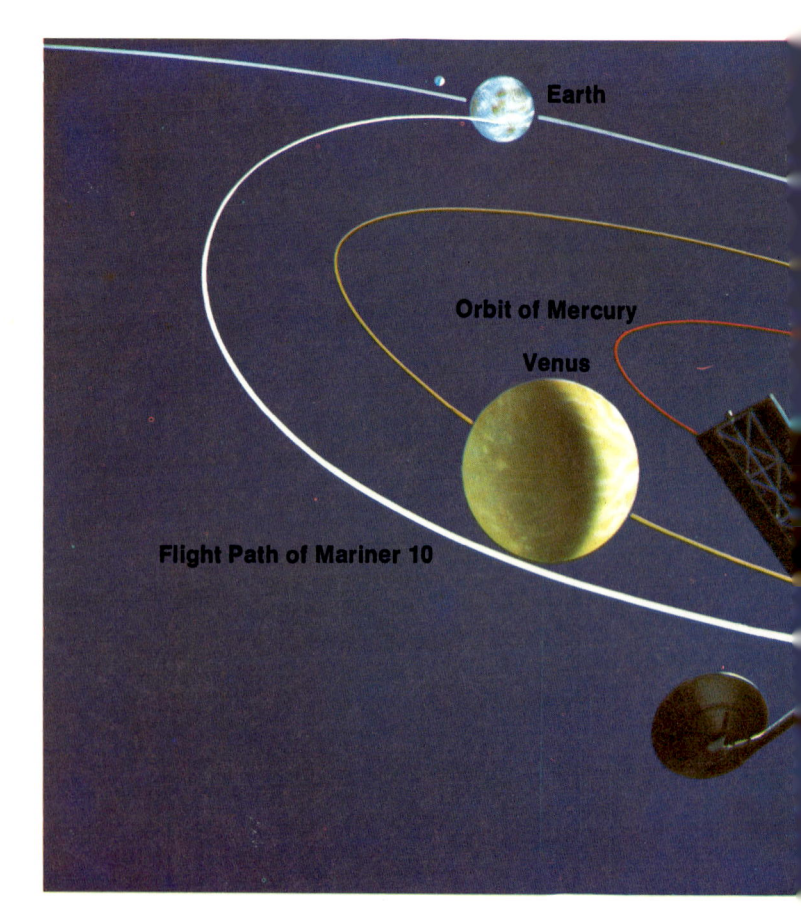

Mariner 10 flew to within 169 miles (270 km) of Mercury's surface. The spacecraft sent back 2,300 very clear pictures. They showed that the surface of Mercury was like the moon's. There were many craters and rocks. It looked like it had been hit by many meteorites.

After *Mariner 10* passed Mercury, it flew by again. It

The flight path of *Mariner 10*. It used the Venus field of gravity to swing toward Mercury.

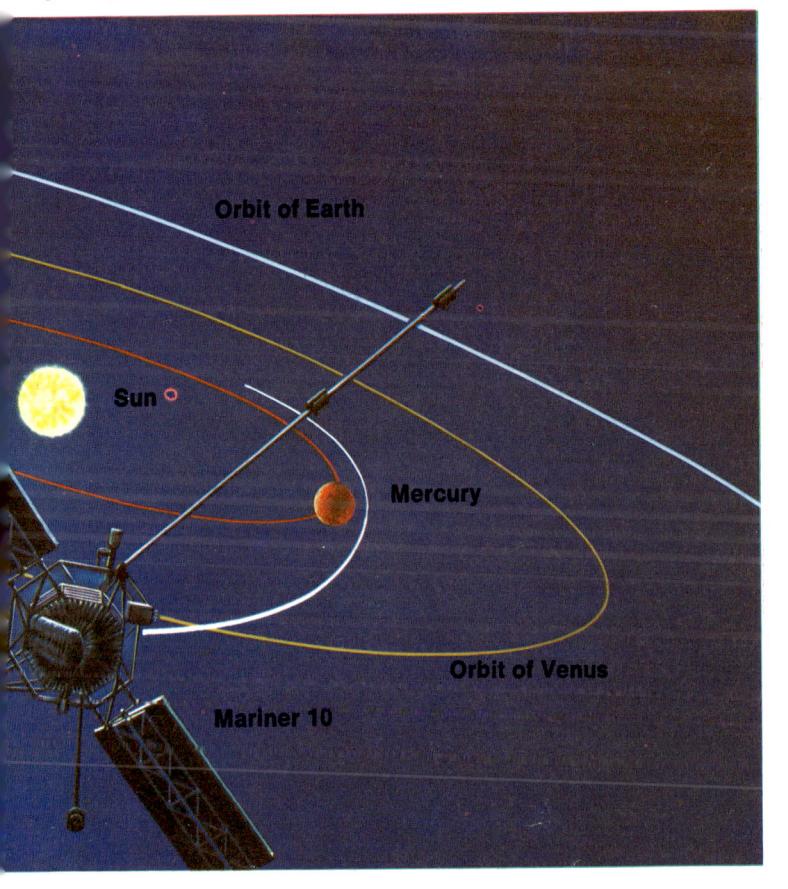

took 1,000 more pictures. One showed a huge crater. It was more than 750 miles (1,200 km) wide.

Mariner 10 went around the sun. Then the spacecraft came near Mercury again. It took more pictures. Altogether, photos were taken of 57 percent of Mercury's surface.

The job was done. The craft shut down and began to fall into the sun.

Cameras of *Mariner 10* found this ring of mountains. It is 812 miles (1,300 km) across.

Space-Age Astronomy

It is not easy to study outer space from the earth. Our atmosphere hides stars and galaxies that are far away. (The atmosphere is the gases around the earth. A galaxy is a group of billions of stars.) And the atmosphere distorts other stars. That means it makes their shapes look different.

But with satellites, scientists can put telescopes out past the atmosphere. Now astronomers can study the huge clouds of gas and dust deep in space. They can study galaxies that are far away.

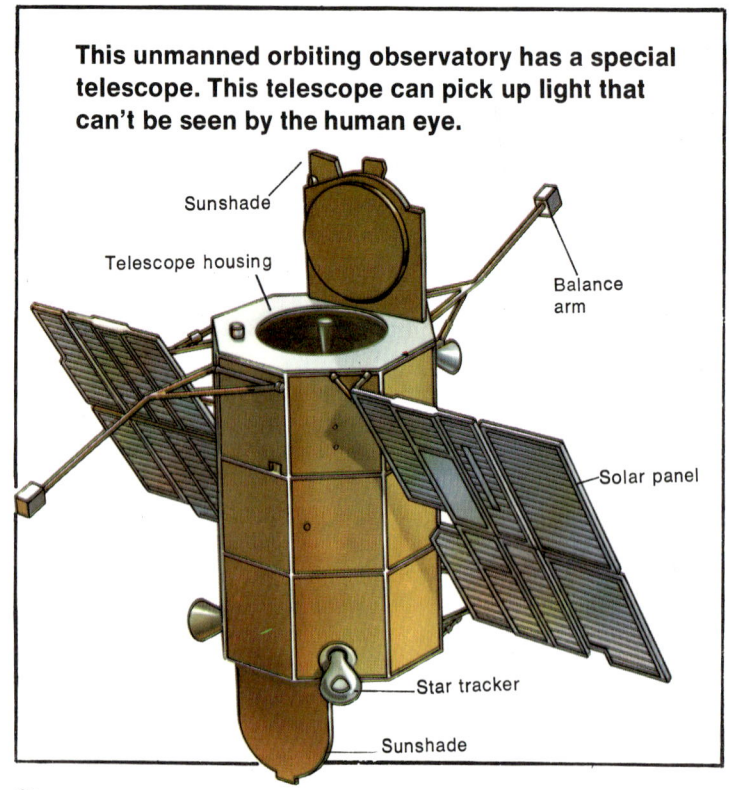

This unmanned orbiting observatory has a special telescope. This telescope can pick up light that can't be seen by the human eye.

Sunshade
Telescope housing
Balance arm
Solar panel
Star tracker
Sunshade

Explorer

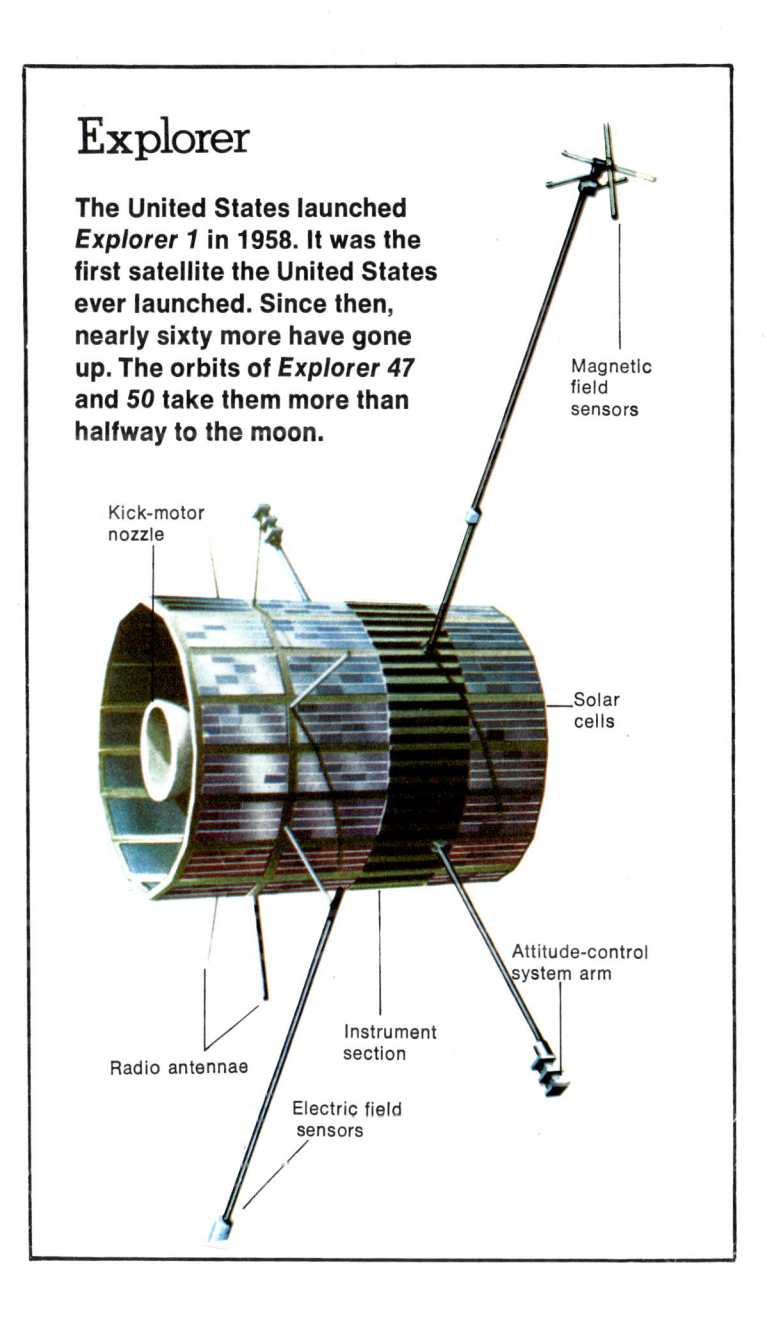

The United States launched *Explorer 1* in 1958. It was the first satellite the United States ever launched. Since then, nearly sixty more have gone up. The orbits of *Explorer 47* and *50* take them more than halfway to the moon.

Magnetic field sensors

Kick-motor nozzle

Solar cells

Attitude-control system arm

Radio antennae

Instrument section

Electric field sensors

The Space Shuttle

The Space Shuttle was designed to fly between the earth and a space station. Other space vehicles can make only one flight. But the Space Shuttle can be used again and again. Scientists plan that the Space Shuttle will make more than 500 flights before 1992.

The main craft, or **orbiter**, looks like most other airplanes. It has a huge fuel tank and two booster rockets. These rockets lift the orbiter into space. Then they drop back to earth by parachute. They can be used again. The Space Shuttle goes into orbit. To return to earth, it flies to the edge of the atmosphere. It lands on a runway like a glider plane.

On April 12, 1981, the *Columbia* Space Shuttle was successfully launched. The first Space Shuttle flight was manned by John W. Young and Robert L. Crippen. The *Columbia* orbited the earth 36 times. It completed some very simple tasks. Crippen and Young tested the ship's equipment. Then the *Columbia* returned safely to earth.

The Shuttle can carry seven people into space. It can carry up to 29 tons of cargo. It will be able to bring more than 14 tons of cargo back to earth.

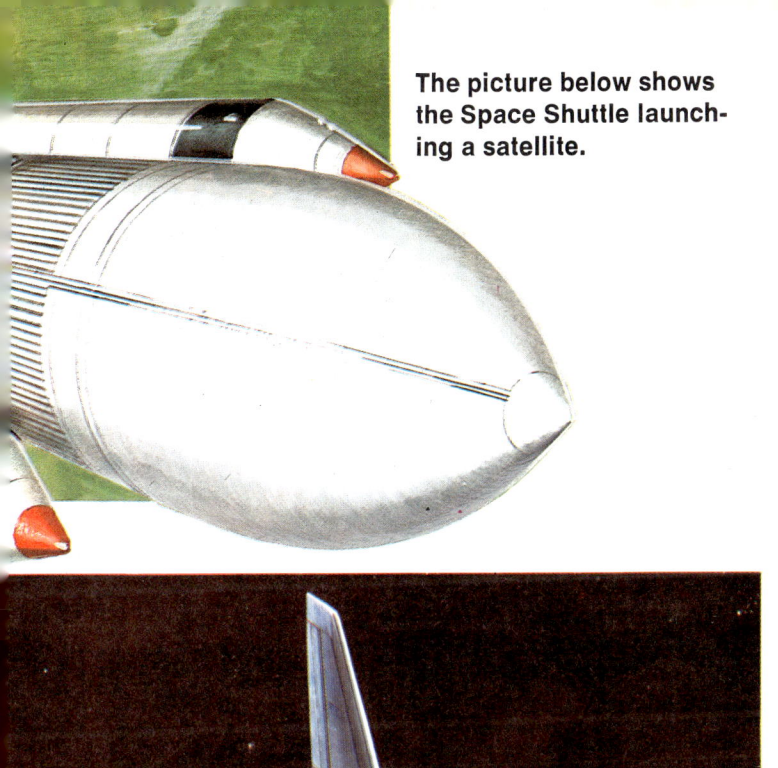

The picture below shows the Space Shuttle launching a satellite.

The cargo of a spacecraft is called its **payload**. The Shuttle has an arm to load and unload its payload. The Space Shuttle can carry large objects such as parts for space stations. It will bring back satellites that need to be repaired.

Launched vehicles and spacecraft made before the Shuttle can carry only trained astronauts. The Shuttle can carry people who are not experienced space travelers. These people will travel in a special passenger compartment. Scientists won't have to spend a long time learning how to travel in space. They will be able to do research in space stations and in outer space.

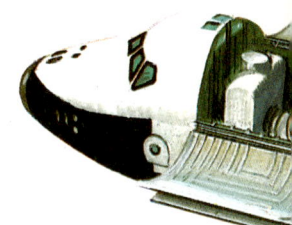

The drawing (right) shows the payload part of the Space Shuttle. The doors are open. A space laboratory is inside. This can be left in orbit. The Shuttle has an arm with a TV camera mounted on the end. It helps the control operator load and unload cargo.

Two booster rockets launch the Shuttle. The rockets parachute back to earth. They will be used again. The Shuttle reaches orbit. It drops the fuel tank. The Shuttle delivers its load. Or, it picks up new cargo. It heads back to earth and glides down to land.

73

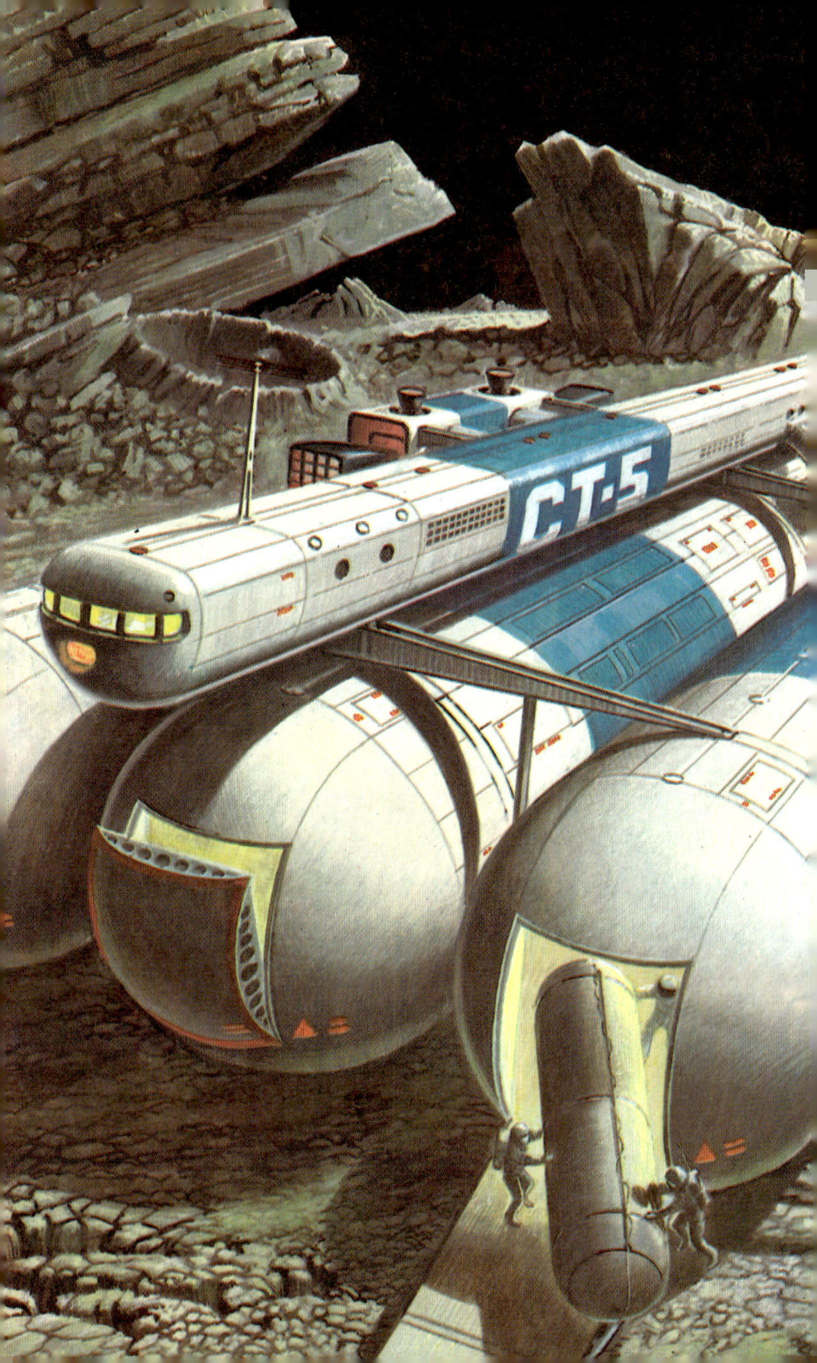
CT-5

What's Next?

We are using up our raw materials on earth. But the other planets have a lot. Someday we may be able to get raw materials from other planets. This picture shows a future base on Phobos. Phobos is a moon of Mars. From Phobos, astronauts could travel near to Mars for exploration.

More Facts

Our Solar System

Our **solar system** is the sun and nine planets and their moons. In addition, there are at least 2,000 **asteroids.** Asteroids are very small.

We do not know how the solar system was made. It may have come from a huge cloud of gas and dust. This cloud was so dense (thick) that it made a field of gravity. Slowly the cloud caved in. Small particles called **atoms** bumped into each other and formed the sun. A ring of objects eventually orbited the sun. They became planets and moons.

Mercury, the nearest planet to the sun, is very hot. Like our moon, it is covered with craters.

Venus, the second planet from the sun, is very hot. Its atmosphere is made up of poisonous carbon dioxide.

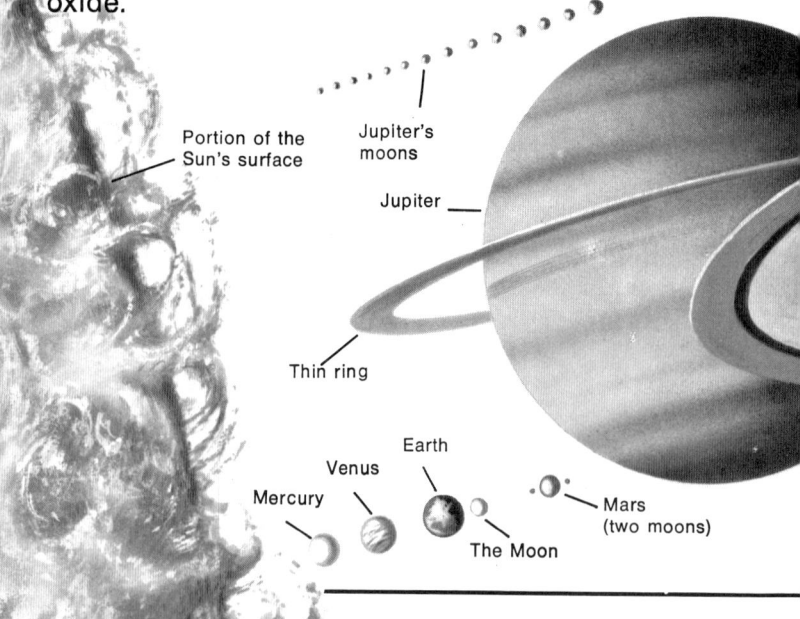

Portion of the Sun's surface

Jupiter's moons

Jupiter

Thin ring

Mercury

Venus

Earth

The Moon

Mars (two moons)

Mars is 37 million miles (60 million km) from the earth. Mars has two moons, Phobos and Deimos.

Jupiter is the next planet. It is huge. Almost all of it is liquid hydrogen. Most of its atmosphere is ammonia and hydrogen. It has sixteen moons.

Saturn has rings of ice and rock. It has at least fifteen moons. Uranus has rings too. It has five moons.

Neptune has two moons. One of its moons is bigger than Mercury.

Pluto is the farthest away. It is almost all ice.

Saturn has at least fifteen moons, the ten largest are shown here

The nine planets of the solar system are of two kinds: (1) heavy rock and metal planets—Mercury, Venus, Earth, and Mars—that are close to the sun; (2) big gas planets—Jupiter, Saturn, Uranus, Neptune. Cold Pluto lies beyond them. It may once have been a moon of Neptune.

Sun and Stars

A **galaxy** is a group of billions of stars. Outer space is full of galaxies. Our galaxy is shaped like a spiral, or coil. It looks like a flat, round dish. Our sun lies near the outside of our galaxy.

There are about 100 billion stars in our galaxy. Scientists have grouped stars that we can see without a telescope or binoculars (field glasses). These stars are grouped

(Above) The position of our sun in the galaxy. (Below) A section of the sun.

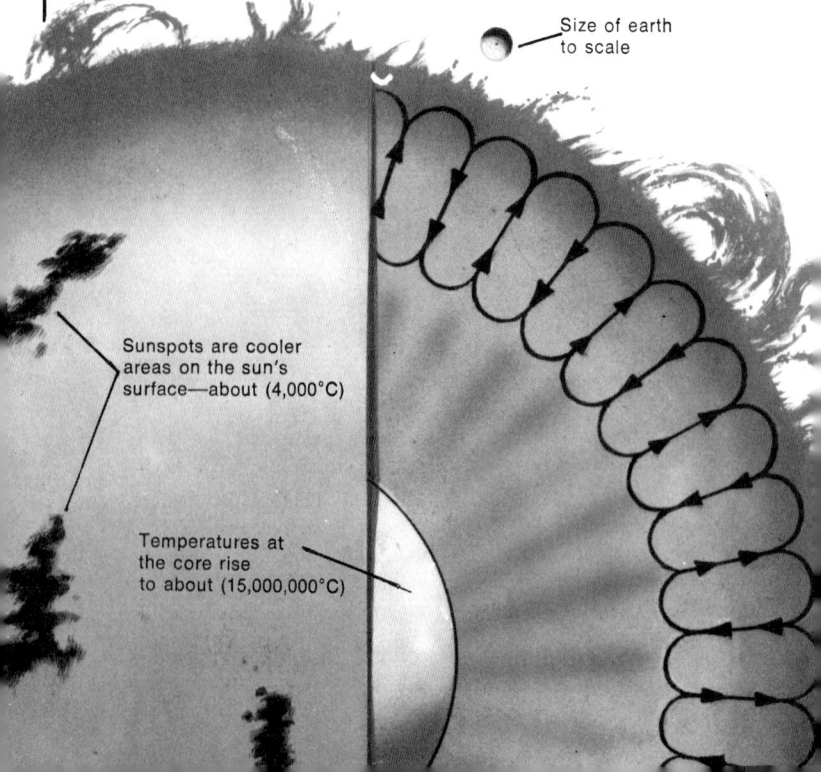

Size of earth to scale

Sunspots are cooler areas on the sun's surface—about (4,000°C)

Temperatures at the core rise to about (15,000,000°C)

according to their size and temperature.

Our own sun is 10,832°F (6,000°C). It is a small star. Our sun burns about 4 million tons of gas every second. That's where its heat comes from. It will take almost 10 billion years for the sun to burn up.

All stars seem to be white, but they are really different colors. The coolest are red. Warmer ones are yellow. White ones are even warmer. The hottest are blue.

Some stars are very big, and others are very small. The sun is 100 times bigger than Sirius B, another star. The star Antares is more than 300 times bigger than our sun.

(Right) This drawing shows that stars come in many sizes. Red Antares is 300 times the size of our sun. Rigel is eighty times the size of our sun. Sirius B is only 24,375 miles (39,000 km) across. It is smaller than the planet Uranus. Wolf 359 is even smaller.

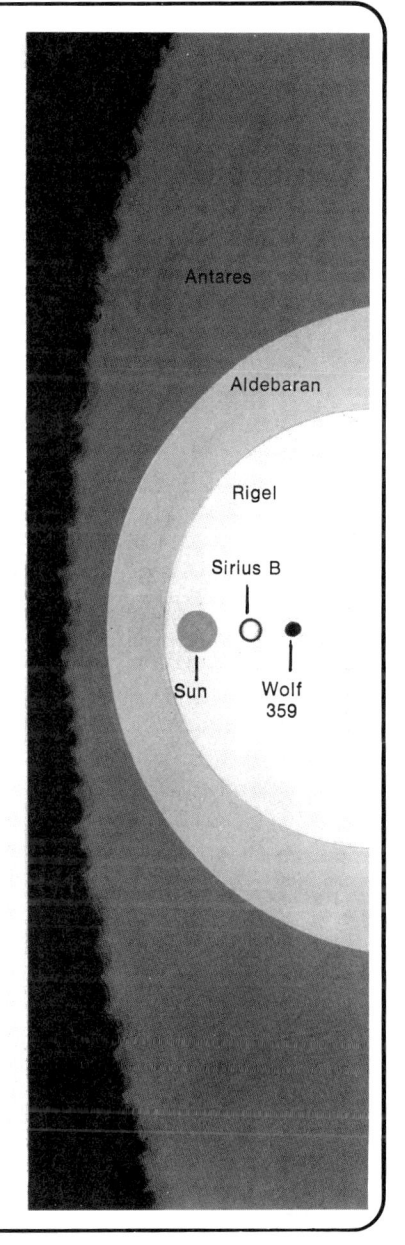

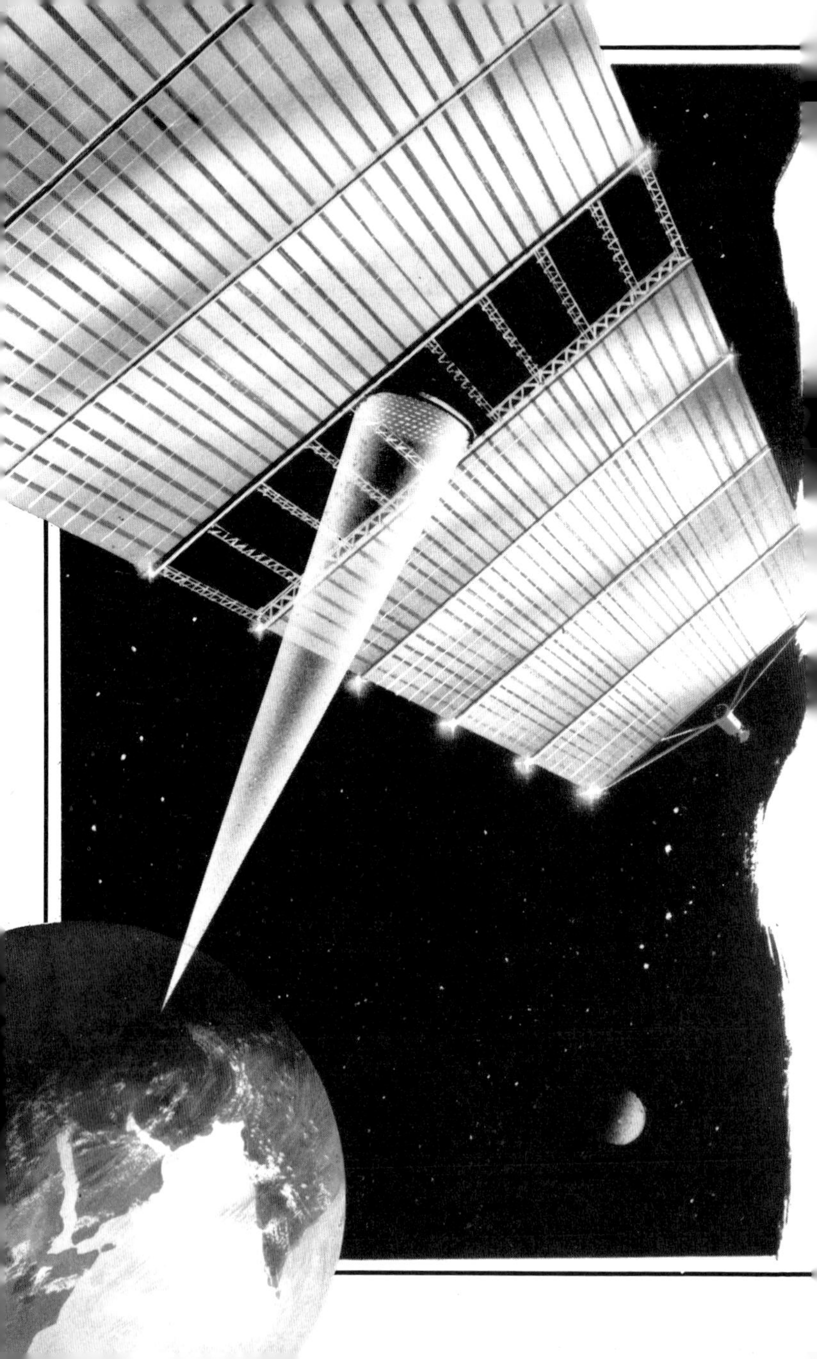

Energy from Space

We use a lot of energy in industry and daily life. Most of this energy comes from fossil fuels. Coal, oil, and natural gas are fossil fuels. But we are using up the fossil fuels on earth. We must find other ways to get energy. Outer space may be the answer.

The sun gives us a lot of energy. Scientists are studying ways to put huge solar panels in orbit around the earth. These panels would change sunlight into energy. This energy would then be changed into microwaves. The microwaves would beam onto the earth. They would be changed into electricity.

(Left) Solar panels in orbit could give us energy. The panels would have to be several miles wide.

(Above) A metal antenna would receive the microwaves from the solar panel. The antenna could be out at sea.

Metals from Space

Scientists plan to build more space stations. But it is very costly to take building materials to space from earth. Scientists are studying how to build with metals that are in space already.

Between 20 and 30 percent of the rock on the moon is metal. Aluminum and iron are two of the metals found in moon rocks. How can we get these metals out? Can they be used for building in space?

Some scientists want to use a "mass driver." It will shoot loads of material from the moon into space.

Ore catchers will take the material to space factories that will process it. To process means to change it into a form that can be used.

A digger picks up rocks from the moon. It takes them to the mass driver. The rocks are shot out into space. A catcher is waiting for them.

(Below right) Material is placed in a tube in a mass driver. There are electromagnets around the tube. This makes a magnetic wave. It acts with the electrical coils to push the material along in the tube. The material is pushed at a very high speed.

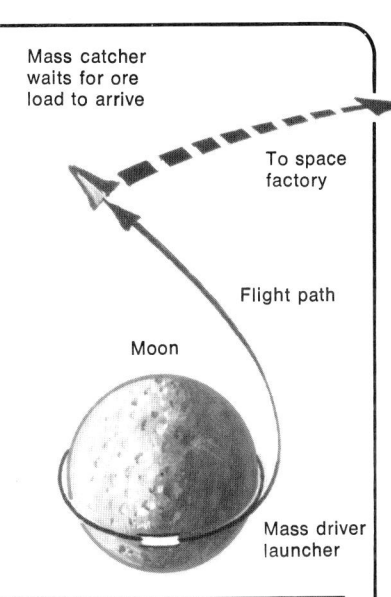

Mass catcher waits for ore load to arrive

To space factory

Flight path

Moon

Mass driver launcher

Nuclear-Powered Spaceships

From 1957 to 1965 the United States worked on a spacecraft that would be powered by **nuclear energy.**

Everything is made of tiny particles called **atoms.** The "glue" that holds the parts of an atom together is atomic, or nuclear, energy. When the **nucleus**, or core, of an atom is split, that energy is released.

In atomic bombs, atoms of uranium or plutonium are split. The nuclear-powered spacecraft would carry thousands of small atomic bombs. One bomb would explode every second. Each ex-

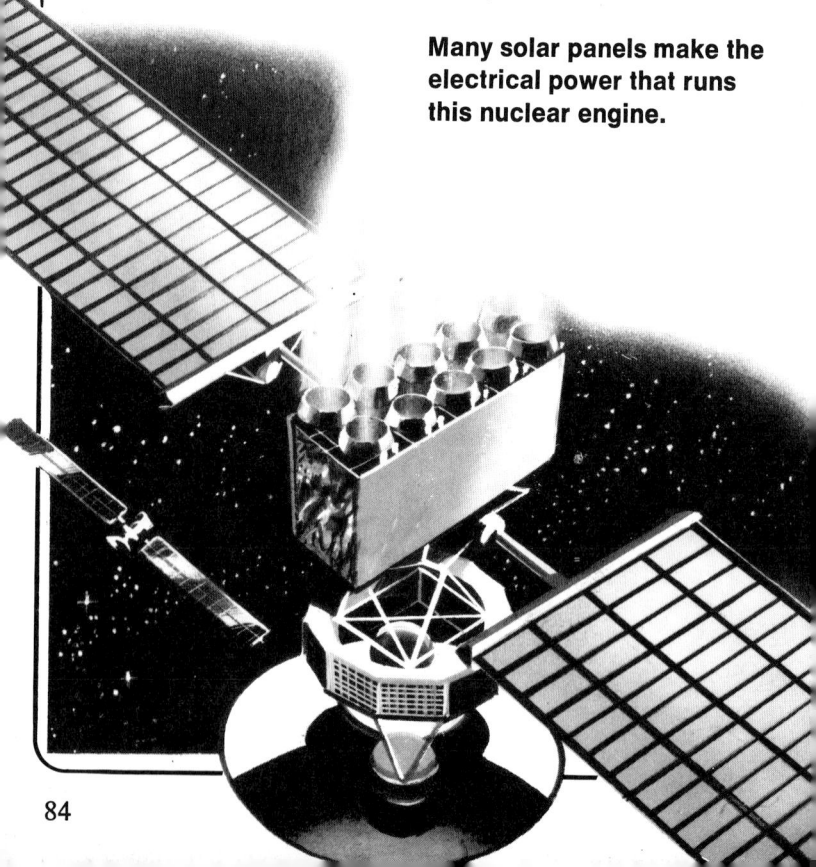

Many solar panels make the electrical power that runs this nuclear engine.

84

plosion would release energy equal to the explosion of 10,000 tons of TNT! (TNT is an explosive used in hand grenades and torpedoes.)

The explosions would make the spacecraft go very fast. Astronauts could go to Mars and back in six months. And, only a small amount of fuel would be needed.

But the waste from nuclear explosions is deadly. Scientists do not know how to get rid of these wastes. And the United States and the Soviet Union agreed to stop certain nuclear tests. So the United States had to drop its plan for a nuclear spacecraft.

The *Orion* spacecraft was never built. It was meant to be powered by nuclear explosions.

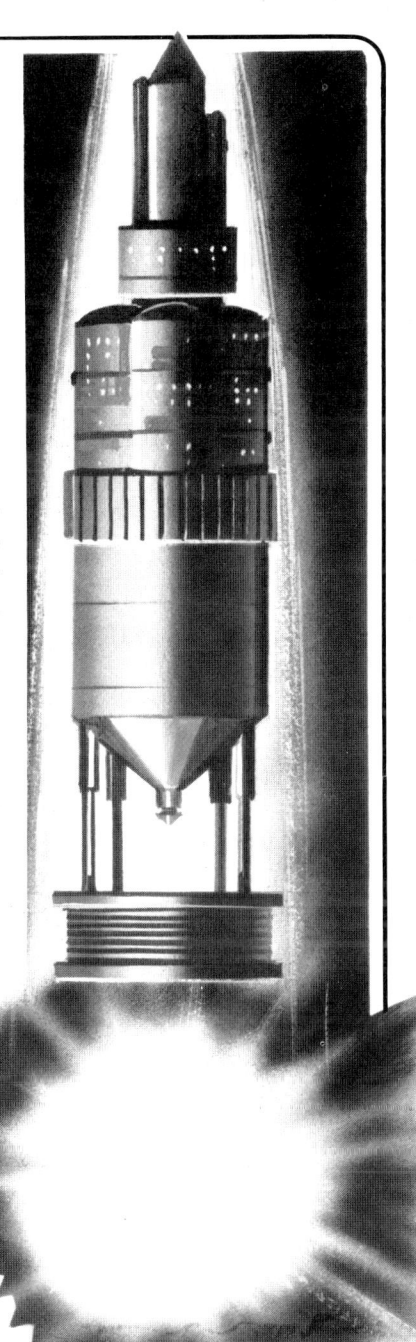

Ramjet

Rockets must carry their fuel with them, so they can't go far. But now scientists are working on the Interstellar *Ramjet.* ("Inter" means "between." "Stellar" means "stars.") It would pick up its own fuel as it went along.

The space between the stars contains hydrogen, helium, and bits of dust. Scientists think these gases might be able to

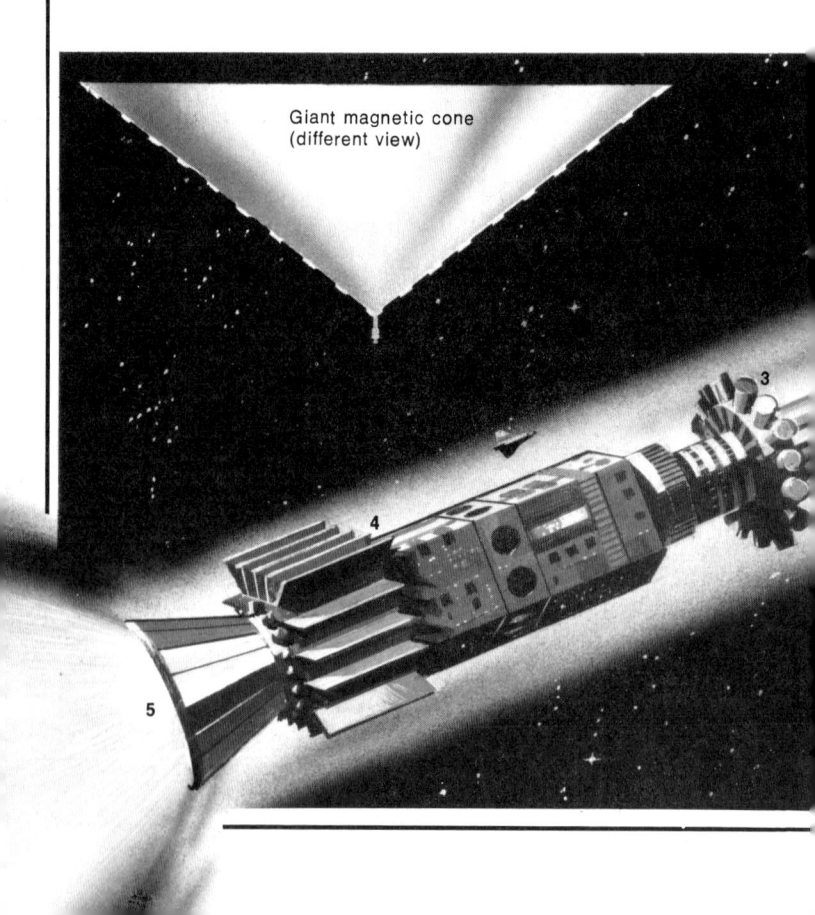

Giant magnetic cone (different view)

provide fuel for a nuclear engine. The *Ramjet*, built like a huge vacuum cleaner, could scoop up the hydrogen. A magnetic cone would "catch" the gas as the craft went through space.

Can a spacecraft collect its own fuel? Scientists think so. But making such a spacecraft will not be easy. And booster rockets will be needed to launch the spacecraft. (1) Magnetic cone. It is 62 to 124 miles (100 to 200 km) wide. (See drawing, upper left.) The cone will sweep up the hydrogen fuel. (2) Ducts carry the hydrogen along. (3) Ring speeds up the gas. (4) Radiators keep the ship's hull (body) cool. (5) Thrust chamber. Here, the gases expand (take up more room). This causes a thrust, or push.

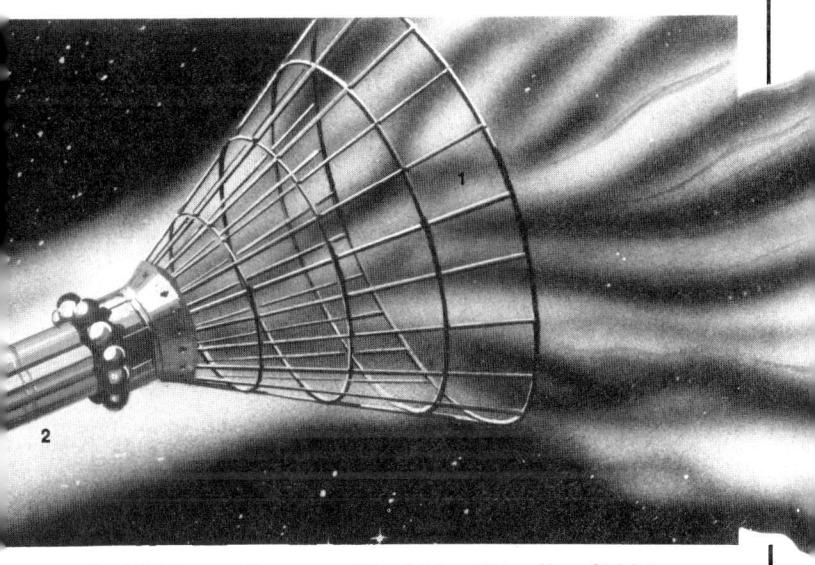

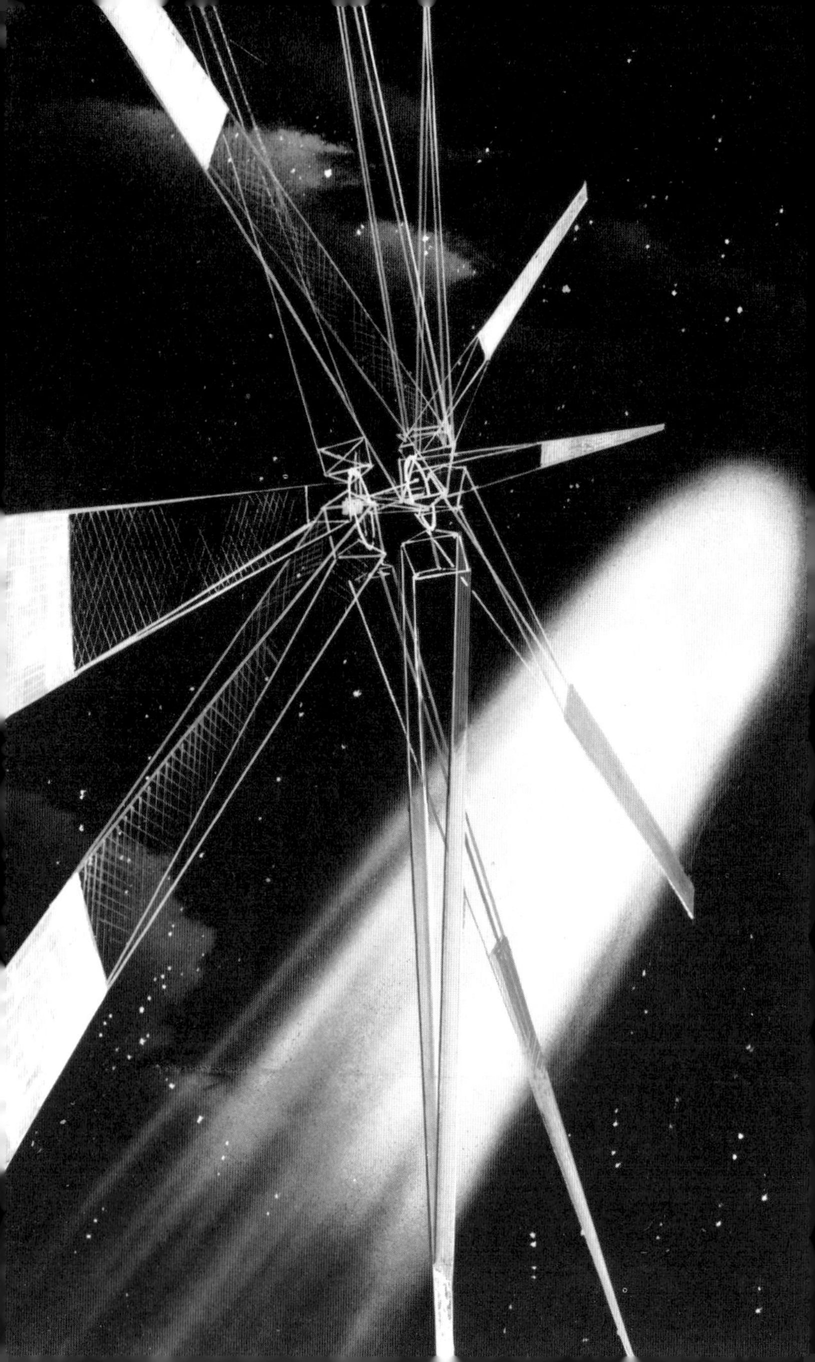

Golden Gate Bridge to scale

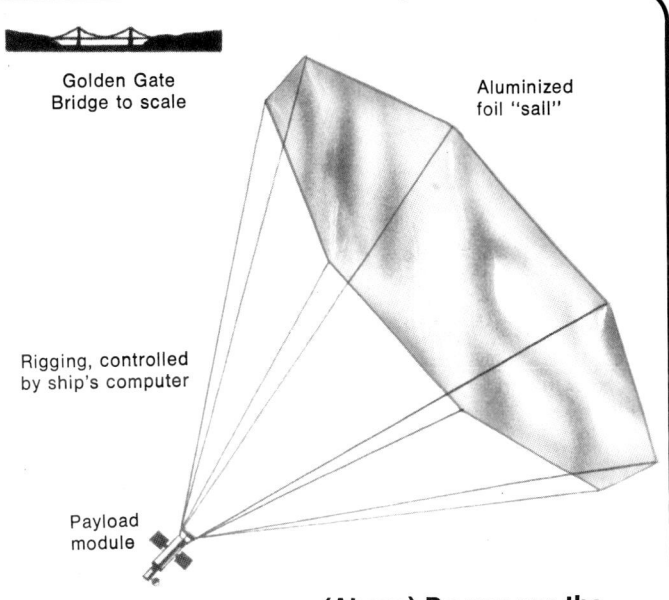

Aluminized foil "sail"

Rigging, controlled by ship's computer

Payload module

(Left) This is a solar sail that's being planned. Each blade would be 4.4 miles (7 km) long.

(Above) Do you see the drawing of the Golden Gate Bridge? That's how little it would seem next to a solar sail.

Sailing on Sunlight

The sun gives off a steady stream of light. This stream presses against anything it its way. A very large, very light object would be pushed faster and faster. Scientists are working on a solar sail that would run on sunlight. It would not need any fuel. This spacecraft would give cheap, fast, and safe flights from one planet to another. The solar sail would be big, but light. It might weigh 5 tons. Its sail could be 0.6 square mile (1 square km). This craft would reach a speed of 2,250 miles (3,600 km) per hour in 11½ days.

Space Firsts

For hundreds of years people have known about rocket-powered flight. But it was not until 1957 that the first satellite bleeped its way around the earth. Since then we have walked on the moon. We have been able to look closely at the other planets. We have sent spacecraft beyond our solar system. We have improved worldwide communications with orbiting satellites.

1957 *Sputnik 1* launched.

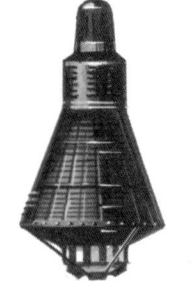

1959 Russian *Lunar 1* missed the moon. It became the first man-made "planet."

1961 Yuri Gagarin was the first person in space. He made one orbit in *Vostok 1* on April 12.

On May 5 Alan Shepard made a spaceflight in *Mercury 5.* He did not go into orbit.

First Russian *Venus* probe launched.

1962 John Glenn was the first United States person in orbit (*Mercury 6*).

1963 Valentina Tereshkova was the first woman in space.

1964 Russians launched *Voskhod 1*. Three men were on board. It made sixteen orbits.

1965 First space walk by Alexei Leonov.

1966 *Luna 9* made first moon landing.

1968 *Apollo 8* flew around moon. (Borman, Lovell, Anders.)

What we have learned has influenced medicine, industry, and even politics. Most of us will never leave the earth. But the Space Age has changed all our lives.

1969 First docking of two manned spacecraft, *Soyuz 4* and *5*.

July 20—Neil Armstrong was first man on the moon. (*Apollo 11*)

1970 *Apollo 13* blasted off with Lovell, Haise, and Swigert. Flight had to be stopped.

September Russians landed *Luna 16* on moon.

1972 United States launched probe, *Pioneer 10*, toward and past Jupiter.

1973 *Skylab 1* launched in May.

1975 *Apollo 18* and *Soyuz 19* docked while orbiting earth.

In October, *Venera 9* and *10* landed on Venus.

1976 In June *Viking 1* landed on Mars. Found no signs of life.

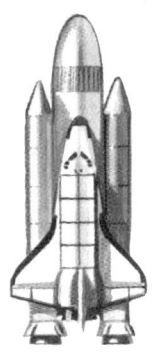

1977 *Soyuz 24* docked in space with *Salyut 5* space laboratory.

U.S. launched *Voyager* craft to Jupiter and Saturn.

1981 First flight of U.S. Space Shuttle.

Index

Acceleration, 36
Agreement in space, between U.S. and Soviet Union, 52–53
Aldrin, Edwin ("Buzz"), Jr., 30, 34, 36
Anders, 29, 90
Animal research, 47–48
Antares, 79
Apollo missions, 11, 17, 25, 27, 28–31, 32–33, 38, 39, 40–43, 44, 46, 90, 91
Apollo spacecraft, 30
Apollo/Soyuz project, 52–53
Armstrong, Neil, 30, 34, 90
Asteroids, 76
Astronauts, 12
Astronomy, 68
Atmosphere, 14, 68
Atomic bombs, 85
Atoms, 76, 84

Basalt, 27
Bean, Alan, 38
Belka, 12
Binoculars, 78
Blood, 34
Bomb, atomic, 85
Booster rocket, 70, 73, 87
Borman, 29, 90
Breccia, 27

Cernan, Eugene, 30, 40, 41
Chemicals on moon, 26
Chinese rocket, 8
Columbia Space Shuttle, 70
Comets, 50
Command module, 29, 30, 32–33
Conrad, 38, 49, 51
Cooper, Gordon, 54
Copernicus crater, 25
Corporal rocket, 10
Cosmonauts, 12, 17

Craters, 24, 25
Crippen, Robert, 70
Cunningham, 29

Deimos, 77
Docking, 18–19, 31
 first, 90
Dogs in space, 7, 12

Eisele, 29
Electromagnet, 83
Energy from space, 81
Escape velocity, 9, 11
EVA (Extra Vehicular Activities) visor, 37
Evans, Ronald, 40
Explorer, 69

Field glasses, 78
First man on the moon, 30
First steps into space, 6–7
Friendship 7, 14
Fuel, 86–87
Future of Space exploration, 75

Gagarin, Yuri, 12, 14, 90
Galaxy, 68, 78
Garriott, Owen, 49
Gemini missions, 12, 15, 17, 54, 56, 57
German rockets, 10
Glenn, John, 14–15, 90
Gordon, 38
Gravity, 8–9, 11, 26
Grissom, Virgil, 14
Ground control, 15

Haise, 39, 91
Handshake in space, 52–53
Heat shield, 13, 14, 46, 47
Highlands of the moon, 25
Humans in space, 12–17

Interplanetary spacecraft, 89
Interstellar *Ramjet*, 86–87

Jupiter, 60–61, 76, 77
Kennedy, John F., 28
Kerwin, Joseph, 49, 51
Kohoutek, 50

Laboratory, space, 72
Laika (first traveler in space), 7
Landing module, 33
Landing on moon, 18, 20, 40–43
Launch vehicle, 10–11
Law of motion, Newton's, 8–9
LEM, 30
Leonov, Alexei, 17, 52, 90
Liberty Bell, 14
Liftoff, 10, 28
Lovell, 29, 39, 90
Luna 1, 90
Luna series, 21, 22, 90–91
LM (Lunar Module), 30–31
Lunar Orbiter, 20–22, 25
 rendezvous, 28
Lunokhod 1, 22

Machines on moon, 22
Magnetic cone, 87
Magnetic waves, 83
Man-made satellite, first, 56
Manned flight to moon, first, 28–30
Mapmaking from space photos, 57
Maria area of moons, 25
Mariner, 62, 64–66, 67
Mars, 76–77, 62, 63
Mars probes, 62
"Mass driver," 82, 83
Mattingly, 39
Mercury, study of, 64–66, 76
Mercury capsule, 14
Mercury probes, 90
Metals from space, 82
Meteorites, 25
Microwaves, 81
Mission Control, 39, 41–42
Module, 18, 33, 39
Moon, exploration of, 20–27; landing, 18–19, 40
Moon rocks, 27
Moon rover, 41–43
Multistage rocket, 10–11

Neptune, 77
Newton's third law of motion, 8–9
Nuclear-powered spacecraft, 84–85

Orbit, 9
Orbiter, 70
Orion spacecraft, 84
Oxidizer, 8

Payload, 72
Peaceful use of space, 52–53
Phobos, 75
Photos, space, 56
Pioneer probes, 60–61, 90–91
Planet, first man-made, 90–91
Planets, 76–77; probing of, 58–69
Pluto, 77
Probe, 58
Problems in spacecraft, 14–15

Ramjet, 86–87
Raw materials from other planets, 75, 82, 83
Red Antares, 79
Red Spot of Jupiter, 61
Remote control TV camera, 43
Retro-rocket, 15
Rigel, 79
Rockets, 8–9; modern, 10–11
RTG, 60

Salyut program, 44–45, 90–91
Satellite, first, 6, 91. man-made, 56
Saturn, 77
Saturn rocket, 11, 28, 32, 40, 50
Schirra, 29
Schmitt, Harrison, 40, 43
Scientists, 56, 72
Seas of moon, 25
Service module, 33
Shepard, Alan, 14, 90
Sirius B, 79
Skylab, 46–49, 50, 51, 91
Soil samples, 63
Solar panels, 81, 84
Solar sail, 89
Solar system, 76–77
Soviet Union, 6, 12
Soyuz missions, 4, 11, 18–19, 44–45, 91
Space age, 90–91
Spacecraft, unmanned, 58
Space firsts, 90–91
Space laboratory, 72
Spaceships, nuclear-powered, 84–85
Space Shuttle, 70–75, 91. *See also* Orbiter Space Shuttle
Space stations, 18–19, 44–45, 46–51
Space suits, 34–37
Sputnik, 6, 7, 90
Stafford, 30, 52
Stages, 10–11
Stars, 78–79
Strelka, 12
Sun (earth's), 78–79 as source of energy, 81, 89
Sunspots, 78
Surveyor series, 3, 22, 23, 38
Swigert, Jack, 39, 91

Telescopes, 46, 68, 78
Tereshkova, Valentina, 13, 90
Tito, Gherman, 12
Traveler in space, first, 7

Unmanned space probes to moon, 20, 58
Uranus, 77

V-2 rockets, 10
Venera probes, 58, 59, 91
Venus, 58, 59, 76–77
Venus probe, 90
Viking probes, 62, 63
Voskhod series, 17, 90
Vostok missions, 12, 13, 19, 90
Voyager program, 61, 91

WAC *Corporal* rocket, 10
Walking in space, 17
Weather predictions, 54, 56
Weight gain during launch, 36
White, Edward, 17
Wolf 359, 79
Woman in space, first, 13

Young, John, 14–15, 30,